RANDALL HROZIENCIK

NOTHING NEW UNDER THE SUN

THE BATTLE OVER ORIGINS FROM THE ANCIENT WORLD TO THE PRESENT

'NOTHING NEW UNDER THE SUN'

The Battle Over Origins

from the

Ancient World to the Present

Randall Hroziencik

'Nothing New Under the Sun'
The Battle Over Origins from the Ancient World to the Present

By

Randall Hroziencik

Published by Athanatos Publishing Group

ISBN: 978-1-64594-054-8

ENDORSEMENTS

Nothing New Under the Sun: The Battle Over Origins from the Ancient World to the Present, by Randall Hroziencik, is a unique convergence of the history concerning the battle for origins and science with its influence on the development of Randy's own Christian worldview. Randall's journey from skepticism to one committed to God's Word and the Gospel of Christ is interwoven into the history and ideas behind the origins debate. As a professing theologian, he has demonstrated a tenacious approach to the study of the history of origins and gives fair apologetic application of the information gleaned, meshing it in a practical way with the past and present views of creation, both his and that of many others. He accomplishes his goal of promoting creation apologetics to equip the saints. The text was helpful to me as a chemist who does not have an in depth theological or historical background, giving a nice overview of the history of creationism, evolutionism, their respective worldview, and the difficult questions of life to which both views attempt to supply answers. The text is full of quotes and personal opinions along with reason and logic, leading the author and reader to the inescapable conclusion that we have a loving Creation-Designer in the Lord Jesus Christ, that His Word is truth and trustworthy from cover to cover, and that the message of the Gospel is relevant to all who were created 'in His image.'

Kenneth W. Funk, Ph.D.

Retired Organic Process Development Chemist and Director of Midwest Creation Fellowship-North in Lindenhurst, Illinois

Since you are holding this book you must be at least considering reading it. So, may I encourage you to devote the few hours required to read it with care? If you do that, you will be challenged to re-examine your ideas on the first three chapters of Genesis, no matter where you start from. In the process you will meet my friend Randy, a man who lives out the direct command of 1 Peter 3:15 to be prepared to answer any faith related question 'with gentleness and respect.'

Reverend Lee Johnson

Retired Senior Pastor of Bethel Baptist Church in Galesburg, Illinois

My friend, Randy Hroziencik, has written a carefully researched work on the history of evolution. It enlightened and stretched me in my thinking. Rooted in the Word of God, this book reveals facts on the history of evolution often unknown. I recommend this book.

Reverend Jim Lacy, D.Min.

Founder of Intentional Interim Church Ministries and Special Appointment Professor of Ministry at Cornerstone University in Grand Rapids, Michigan

I heartily recommend Randall Hroziencik's second book *Nothing New Under the Sun*. Truly, as Solomon said, there is nothing new under the sun, and this applies especially to the young earth versus old earth debate. But, actually, perhaps there is something new under the sun, which is a passionate, fair, thorough, well-researched, and a theologically based treatment of the question which gives voice to the wide range of Christian views of origins. It is a good sign that Randall gets push-back from both 'sides' in the debate. We need to continue to listen, especially to someone like Randall who will passionately pursue all avenues of research and give voice to the widest possible range of faithful believers in the one true Creator of all.

John Oakes, Ph.D.

President of the Apologetics Research Society and author of *Is There a God? Questions about Science and the Bible* and *Reasons for Belief: A Handbook of Christian Evidence*

This could have easily ended up as another impersonal apologetics book of 'white noise,' or even just a historical addendum to apologetics in general. Thankfully, Randy's personal touches and his willingness to explore the implications laid out by his historical recounting makes this a much more compelling and impactful work.

Reverend Weston Oxley

Pastor of Adult Ministries at Bethel Baptist Church in Galesburg, Illinois

Randy Hroziencik's second book is a well-researched examination of the current state of the evolution/creation debate. He begins with some little-known history and then brings it up to today. 1 Thessalonians 5:21 says to, "test everything; hold fast what is good." *Nothing New Under the Sun* will help you to test 'everything' about the subject of Origins – and then help you decide what is good and true.

Helmut Welke

Ambassador for Logos Research Associates and Founder of the Quad City Creation Science Association

Randy is a humble blessing to the church and all those with a hunger for truth. His latest book, *Nothing New Under the Sun*, is a masterful exploration of the history and concepts concerning origins. He has taken complicated philosophical, historical, and theological concepts and presented them for his readers consideration with personal and relevant insights. People from all walks of life will benefit from his scholarly work, and those unfamiliar with apologetics will be encouraged to thoughtfully examine their worldview. His love for God's Word and commitment to the Gospel of Christ miraculously shines through in a season where many Christians do not prioritize thoughtful examinations of their own beliefs. I'm thankful for the way God has used Randy as a blessing to our local church and what his writings will do for others beyond our community for years to come.

Reverend Scott Wilson

Senior Pastor of Bethel Baptist Church in Galesburg, Illinois

DEDICATION

This book is dedicated to my late brother-in-law, Francis E. 'Frank' Polite (November 9, 1955-August 21, 2020). Frank was more than just a brother-in-law – he was simply my brother. I learned a lot from him over the 45 years that we knew each other. Unfortunately for Frank, he had to put up with a weird brother-in-law (me) who surely drove him to near-insanity a time or two, especially during my pre-teen and teenage years!

Frank had to battle serious health issues for much of his adult life, first cancer and then a serious heart condition which was undoubtedly caused by his cancer treatment. I miss him greatly, as does everyone who ever knew him. Until we meet again.

"For as in Adam all die, so in Christ all will be made alive."

Paul's First Letter to the Corinthians (15:22)

"And I, when I am lifted up from the earth, will draw all people to myself."

John's Gospel (12:32)

"My Father's house has many rooms; if that were not so, would I have told you that I am going there to prepare a place for you?"

John's Gospel (14:2)

CONTENTS

ACKNOWLEDGMENTS

There have been many excellent books written on the history of the origins debate (evolutionism versus creationism). So why another? There are several reasons for my writing this book. First, I love the topic! For several years I have been examining the history of evolutionary thinking, to determine if it is based solely upon scientific discoveries in the last few centuries or is instead a much older philosophy intended to supplant God's revelation concerning origins. I decided to turn my research into a book, to share with others what I have discovered.

Second, I wanted to combine the best of each resource I used in my research. Most books or papers focused only on one period of history. I discovered that very few resources concentrated on evolutionism in the ancient world, prior to the time of Christ, and fewer still addressed the possibility of evolutionism in the pre-Flood or early post-Flood world. Likewise, some resources spent more time discussing evolutionism in the modern era at the expense of the Medieval, Renaissance, and Enlightenment eras. That left me wanting to combine the best of each book or paper used into one resource, covering the ancient world to the present.

Third, I wanted to stress the spiritual warfare aspect of the origins debate. Some resources did this very well, especially Henry Morris' *The Long War Against God*, whereas others never really mentioned it. The belief that evolutionism is an ancient 'doctrine of demons' intended to lead people astray is a vitally important issue, however, and we will all do well to understand it.

Fourth, as much as I love studying creationism and apologetics, I consider myself a theologian who is every bit as concerned with proclaiming doctrine as I am defending the faith. Many unbelievers claim that their main reason for denying Christianity revolves around the belief that 'modern' science dispels the need for God. Although I have no doubt that the supposed 'problem of science' ranks at the top of objections for some skeptics, my experience has led me to conclude that skeptical objections are more often concerned with theology. Among those theological objections to the Christian faith, the problem of suffering and evil and salvation in Christ alone seem to be at the top of the list. Therefore, I wanted to address these issues in depth as well.

Fifth, I wanted to offer a resource with lots of suggestions for further study, since not many books or papers on this topic do that. I could revise and expand this book for as long as I live, as one can always dig deeper into the history, philosophy, and science of this fascinating topic. But the important thing is to understand the basics, and that can be accomplished in a book such as this. That is the reason why I offer a list of resources for further study at the back of the book, as well as providing several resource recommendations throughout the text itself. As you go through this study you will likely find topics that you would like to examine in greater detail. When you do, it is my intention to get you headed in the right direction.

Finally, and this is the most personal reason by far, I have spent the past few years coming to terms with my mortality. Intellectually, I always knew that someday I will die. Everyone does, of course. But 'head knowledge' about death is different from almost experiencing it first-hand. At the tail-end of 2018 I almost wrapped up my career as a human being. As a result of my brush with death, I have become much more focused on the 'big questions of life.' That is saying a lot, for I am an ordained seminary graduate with experience in apologetic speaking and writing. I have also worked with cancer patients for over thirty years, and I have officiated many funerals. I was already thinking about existential issues a lot before my heart attack, but now it borders on ridiculous! As a result of my experience, I feel the need to leave something behind for my family and friends – something of real value and not just material things. I hope that this book will find its way into the hands of my family, especially my children, grandchildren, great-grandchildren, and so on down the line. The information contained in this book can make all the difference in the world – and all the difference in forming a worldview.

My Heartfelt Gratitude

I would like to acknowledge those who have inspired and helped me in this endeavor. Almost every author stands upon the shoulders of those who came before him or her, and that is certainly the case with me. Each one of the authors listed in the resources for further study inspired me to write this book. They are people committed to both serving the Church and reaching outside its' walls with the message of creation and the Gospel.

When writing this book, I sought the input of several people who I consider to be extremely knowledgeable in theology and science. Since I asked people from a variety of theological viewpoints, not everyone agrees on the topics contained within the pages of this book. I wanted to take the 'Abraham Lincoln' approach to seeking the advice of others. It has been written that Lincoln surrounded himself with politicians who did not exactly see eye-to-eye with him on various issues. Lincoln knew that by doing this, there would be a better system of checks and balances in place. Some of my friends are young earth creationists, and some are old earth creationists, while I have been both at various times. This approach has resulted in a book that is more informative regarding many topics. It is unlikely that anyone will agree with me 100% on everything written, but I have made it a point to explain why I believe what I believe. In the end, that is the best anyone can do.

I would like to thank several people for their willingness to read through an early manuscript of this book, and for taking the time to offer me numerous suggestions. As they are all equally important to me, I have decided to list them in the order that they got back to me with their suggestions. (It was either that or the always overdone alphabetical order by last name.)

Roger Brown, an elder at my home church, is a tireless leader in many ways and in many areas. Roger told me after reading an early draft of the book that it was not only more reader-friendly than my first book, but it also more closely reflected my classroom teaching style. Therefore, as I adjusted the text in later draft's I always kept those positive comments in mind. The result has been, I believe, a book that has remained more reader-friendly than *Worldviews in Collision*, and it also more closely reflects my classroom teaching style. Thank you, Roger, for that helpful advice.

Lee Johnson, the retired senior pastor of Bethel Baptist Church in Galesburg, Illinois, is the most amazing Bible teacher I know – and I have sat under the tutelage of quite a few Bible teachers in my twenty-five plus years as a believer. As I mentioned in the dedication page of my first book, despite completing formal seminary degrees I really received my theological education from Pastor Lee. He always challenged me to think, and oftentimes re-think, my position on certain doctrines of the faith, which has made me a better, more-rounded theologian and apologist. I greatly appreciate Pastor Lee's willingness to write the Afterward as well. His input has undoubtedly made this a better, more thought-out book, and I greatly value his wisdom, guidance, and friendship.

Helmut Welke, the founding president of the Quad City Creation Science Association, is a gifted speaker who is extremely knowledgeable in every area of creation science. I have had the pleasure of meeting several of the top creationist speakers in the world today, yet I am convinced that Helmut is at the top of the list when it comes to combining knowledge about creationism with the ability to impart that knowledge to others. Not everyone who is knowledgeable is necessarily an effective teacher, but Helmut is one of those rare individuals. I am truly blessed by having Helmut live so close by, and for his willingness to read through several drafts of this book and offer numerous suggestions. Although everyone who helped me in this endeavor is important, I must say that Helmut spent more time on this project than anyone else by far. He also wrote a very thought-provoking Foreward as well. I quote Helmut extensively in chapter six, which focuses on the age of the creation. This book would not be nearly as accurate without his input.

Weston Oxley, Pastor of Adult Ministries at Bethel Baptist Church in Galesburg, Illinois, is someone whose opinion I value deeply. Weston and I think alike: Our interests are similar, and we tend to approach the deep issues of theology in the same way. Not that we always agree on everything, but our process of thinking through the issues tends to be remarkably similar. This partly explains why I value Weston's opinion so greatly. Weston gave me some great suggestions, which helped to make the book more reader friendly. Thank you, Weston, for taking the time to sit down with me and go through your suggestions.

Dr. Kenneth W. Funk, a retired organic chemist, is the Director of Midwest Creation Fellowship-North in Lindenhurst, Illinois. I did not have the pleasure of becoming acquainted with Ken until the very end of the writing process, but I thank God he entered the picture when he did and not after I completed the book. (Thank you, Helmut, for introducing us.) Ken offered numerous suggestions, in addition to providing me with a healthy dose of encouragement. As with the others, this book would not be nearly as accurate without his input.

I consider Dr. John Oakes as being one of my primary mentors. John was my instructor during the year and a half that I took to complete the Apologetics Research Society's Certificate in Christian Apologetics. He is, without a doubt in my mind, the 'best of the best' among apologists today. Although his doctorate is in the sciences, he is amazingly adept in theology, history, philosophy, and world religions as well. That is a rare combination, indeed. I knew there would be no way that I could undertake this project without his input. John provided me with many great suggestions, which served to make this book more accurate. Thank you, John, for taking the time to assist me in this project.

I was blessed to meet Dr. Jim Lacy a few years ago, when he became our interim senior pastor during Scott Wilson's short-term deployment to the Middle East. (Which probably felt like a million years to Pastor Scott!) In addition to serving various churches as an interim pastor, Pastor Jim is also a professor of ministry at Cornerstone University in Grand Rapids, Michigan. We became good friends during his time spent in Galesburg, Illinois. While writing this book I knew I needed his input as a pastor, professor, and friend. Some people who are old earth creationists were probably a bit disappointed by my position, and other friends who are young earth creationists must have felt that I was being too soft toward the other side. Pastor Jim, however, recognized my approach as being both fair and balanced, so I knew I was onto something right. Thank you, Jim, for taking the time to help me.

When writing the early drafts of this book, I had the best-of-the-best checking my work. If any errors remain, however, I take full responsibility for them.

Finally, I would like to thank my family for their continuing support. My wife, children, and grandchildren mean everything to me, as do the rest of my extended family. As already mentioned, I hope that this book finds its way into the hands of not only my children and grandchildren, but even later generations as well. That would be incredible! Our immediate family consists of our son Chris and his wife Nicole, and our daughter Heather and her husband Anthony, as well as our four granddaughters Alexa, Eleanor, Jane, and Maisie. Love you all.

FOREWORD

Helmut Welke, MS

Retired Engineering Manager & Founder of the Quad City Creation Science Association

"I love to think of nature as an unlimited broadcasting station, through which God speaks to us every hour, if we will only tune in."—George Washington Carver (1864-1943)

Professor Carver was a great researcher and teacher in the early twentieth century (MS degree, 1896 from Iowa State University). And I love his quote. It tells us two very interesting things. First, he says that nature broadcasts what many consider to be an obvious conclusion: that God has revealed Himself in His creation. (Paul makes this plain in Romans 1:20). Second, there is a caveat here, a condition: "if we will only tune in." We do need to listen and understand. Learning to "tune in" in order to understand and make sense of what we observe in the world is not always easy. Discovering the answers to the questions that nature can answer is often a struggle, for a variety of reasons.

One reason might be that we don't have the time, or we just don't want to listen and see the obvious God message. Another reason may be a lack of training to see the answers that nature can provide – as well as those it cannot provide. But a final and major reason is the 'noise' around us. There are other 'broadcasts' we can pick up that are saying things that are in conflict with each other, with God's message, and may often make it all seem like so much noise. But the effort to 'tune-in' and get the message that God wants us to hear is worth the effort. To understand the struggles that many have about 'tuning in' and getting the 'God message' (apart from the other noise), is what Randy Hroziencik's new book is about.

Randy's book is well-written. He includes ideas and conclusions from the struggles of others, and how he has determined the good broadcasts from the not-so-good. He includes numerous footnotes and ideas for further research. His style is almost like a conversation. He writes excellent prose that is clear and intelligent, but also very understandable and easy to read. He includes his own struggles with 'tuning in,' sorting through the various broadcasts that have been there for many generations. Since he explains his own various attempts to understand both the science and the Scripture, this book reads at times like a novel – one that is hard to put down.

Do we need another book about the origins of nature and the debate between Creation and evolution? Well, this book is a 'yes.' It has three major parts. Beginning with the Introduction and through chapter four, Randy takes us through the history of the theory of evolution and how there is 'nothing new under the sun.' He walks us through some historical narratives and philosophies and the various ideas that are in clear opposition to the biblical account of God creating the universe, the world, and making human beings from scratch – as opposed to evolving over time from a chimp-like ancestor. This first section ends with his chapter four discussion on the modern Intelligent Design movement. He provides some good background on why this is an attractive movement for many people of different worldviews, simply because of the sciences of microbiology and biochemistry. This is the broadcast that nature provides, showing us that there really is an intelligent designer.

The second section, composed of chapters five and six, provide a summary of the scientific reasons that the evolutionary broadcast has serious problems, if not strong reasons to discard Darwinism altogether. Here again, Randy provides good summaries with just enough details to help the reader understand the major points and perhaps want even more. As an engineer, this is the area that always interests me the most. I have always liked science, and early on I developed enough of a skeptical attitude to want to see the actual evidence – not just the so-called accepted interpretation. It is amazing what you can find by asking good questions, such as "may I see the actual fossil, please?" and not just the artistic renditions and accompanying story line. Or, "how do you know that rock or fossil is that old? What methods were used to determine that number, and what are the assumptions behind the final arithmetic?" Again, it is amazing what you can learn by tuning in to a 'broadcast' to get a few details and determine for yourself if the given conclusions are warranted.

Chapter five takes a good look at origins, both the origin of life and the universe itself. Again, Randy provides good summaries of the various arguments without going too far into the weeds. He also looks at the complexity of the DNA molecule, but then also the complexity of the human mind. This section has some good quotes and an interesting conclusion.

Chapter six goes into the thorny issue of the age of the earth and the universe. Randy does a fine job explaining his own views, but gives decent explanations for the views of others, both old and young ages. He discusses radiometric dating techniques but then goes into the related theological issue of death before sin. This discussion is very good, and again provides some excellent food for thought, wherever your current viewpoint may lie. There is even more in this chapter, and I think that you will enjoy the discussion Randy brings with the different viewpoints – but also with some of the modern research that is discussed.

The last section (chapters seven through nine) continues by tackling more theological issues such as, "Why is there suffering and death in this world?" and "What about those who never heard the name of Jesus?" He makes a final case concerning resurrection and the afterlife in chapter nine. This is what it all comes to: How will you spend eternity?

What makes Randy's book stand out is the historical content and the concise clarity of different points of view. It is a good, enjoyable, and enlightening read– helping you discern the many 'broadcasts' we hear in our modern world. I expect you will enjoy it.

Helmut Welke; December 2020. Contact: Info@QCCSA.org

"The grass withers, the flower fades, but the Word of our God stands forever." (Isaiah 40:8)

INTRODUCTION

Evolutionism: An Ancient 'Doctrine of Demons'

When reading this book, you might be surprised to discover that evolutionism is not a new idea.[1] It is a very ancient philosophical belief that is perhaps as old as humanity itself. I have encountered many people who are surprised to learn that evolutionary teachings did not begin with Charles Darwin in the middle of the nineteenth century but rather go back well into the ancient world, even beyond the time of the Greeks.

There are many people today who would be quite surprised at the notion that evolutionism should be questioned, let alone rejected, by anyone. Many years ago, I was one of those people. As a religious skeptic and an evolutionist, with a few 'New Age' ideas mixed in, the notion that molecules-to-man evolutionism should be seriously contested was outrageous to me.[2] After all, everyone knows that evolutionism is true. What kind of bonehead would even dare question it? Not me. I would never fall for that type of nonsense![3]

The problem I had was not intellectual, however. Rather, it was spiritual. The Apostle Paul described my condition better than anyone else ever has, or ever will: "The person without the Spirit does not accept the things that come from the Spirit of God but considers them foolishness, and cannot understand them because they are discerned only through the Spirit."[4] Without the regeneration of the heart and mind that comes from accepting Christ as Lord and Savior, I could never begin to understand the truth about God and origins. When I accepted Christ for who he really is – the Creator of everything, and the Redeemer of fallen man – my thinking was changed, and I began to see God for who he says he is in the pages of Scripture. Paul, once again, points this out better than anyone: "What we have received is not the spirit of the world, but the Spirit who is from God, so that we may understand what God has freely given us."[5] Once I became a follower of Christ, my evolutionary beliefs were seriously challenged. In this book I will share some of the insights into the history and ideas behind the origins debate that I discovered along the way.

---

[1] The term 'evolutionism' refers to cosmic, geological, and biological change over time.

[2] Even as a skeptic, when it came to answering the big questions of life, atheism had no explanatory power for me whatsoever.

[3] Many people today will be quick to think like I once did, saying to themselves and others, "Everyone with a brain knows that evolutionism is true!" But how many people have taken the time to investigate the issue? In my experience, very few. Most people cling to the tired, worn out ideas of secularism without even realizing it.

[4] 1 Corinthians 2:14, NIV.

[5] 1 Corinthians 2:12, NIV.

'Doctrines of Demons'

Paul wrote these words to his young protégé Timothy: "Now the Spirit expressly says that in later times some will depart from the faith by devoting themselves to deceitful spirits and teachings of demons."[6] Some versions of Scripture use the word 'doctrines' rather than 'teachings,'[7] but what exactly is a 'doctrine of demons'? It is nothing more than any false idea that is used by Satan and his angels to prevent a person from coming to the truth that is revealed in Scripture. It goes without saying that there are many such doctrines or ideas. Based upon the spiritual dimension of the evolutionism-creationism conflict,[8] it seems clear that evolutionary teaching is one of those doctrines.

God has revealed to us the reality of a spiritual conflict that we, as believers, wage against evil forces. We call that conflict 'spiritual warfare.' Paul addressed this ongoing battle in his letter to the Ephesian churches:

> Finally, be strong in the Lord and in his mighty power. Put on the full armor of God, so that you can take your stand against the devil's schemes. For our struggle is not against flesh and blood, but against the rulers, against the authorities, against the powers of this dark world and against the spiritual forces of evil in the heavenly realms. Therefore put on the full armor of God, so that when the day of evil comes, you may be able to stand your ground, and after you have done everything, to stand. Stand firm then, with the belt of truth buckled around your waist, with the breastplate of righteousness in place, and with your feet fitted with the readiness that comes from the gospel of peace. In addition to all this, take up the shield of faith, with which you can extinguish all the flaming arrows of the evil one. Take the helmet of salvation and the sword of the Spirit, which is the word of God.[9]

The unbelieving world laughs at the idea of Satan as a real being. Satan is only real for as long as it takes to watch *The Exorcist* or some other concoction from Hollywood. Beyond that, skeptics consider the idea of an evil, non-human enemy of God to be the stuff of legend – and nothing more. Henry Morris, one of the founders of the modern creation science movement, noted the rising prevalence of Satanism and the occult in modern times: "Even in this supposed

---

[6] 1 Timothy 4:1, ESV.

[7] For instance, the KJV uses the term 'doctrines of devils,' while the NKJV uses the term 'doctrines of demons.'

[8] The evolutionism-creationism conflict is clearly spiritual in nature (1 Corinthians 2:14). Based upon Romans 1:20, Romans 2:14-15, and Ecclesiastes 3:11 all people inherently recognize the Creator, and are therefore 'without excuse' when denying God.

[9] Ephesians 6:10-17, NIV.

scientific age, occultism and Satanism are probably followed by more people today than ever before in history."[10] Even though Morris wrote this twenty years ago, it is still true today. Despite the claims of skeptics, throughout history mankind has widely accepted the idea of a malevolent spirit-being who is in constant conflict with God. Many of the world's religions, such as Hinduism, Buddhism, Zoroastrianism, and Islam, discuss their version of the 'Evil One.' According to Revelation 13:4, there will even come a time in the future when people will openly worship Satan in great numbers, although I contend that many people in the world are doing that now (although unwittingly) through their acceptance of occultic ideas and 'idolatry unaware.' We will explore the nature, and motivation, of Satan in more detail in chapter one.

This spiritual warfare constantly plays itself out in the origins debate. The Renaissance philosopher Robert Fludd, despite being enthralled with the pagan ideas of "astrology, alchemy, Christian cabbalism, and Neo-Platonism,"[11] recognized that most of the ancient Greek writings were "corrupt and diabolical being founded on the world's wisdom, and therefore sourced from the 'Prince of Darkness.'"[12] Like Fludd, we must recognize the spiritual side of the origins conflict, for it is very real indeed.

Paul, in his letter to the Church in Rome, described how the fall of man negatively affected our view of God and creation (Romans 1:18-25). Paul tells us that, from the time of the fall of man to the present, sinful man naturally tends toward a wrong view of our Creator (Romans 3:23; 1 Corinthians 2:14). Evolutionism is a way Satan keeps fallen man away from the truth of biblical creationism (2 Corinthians 4:4), which in turn can lead to a rejection of the Gospel message.[13]

In modern times there has been a strong push worldwide for people to seek knowledge of what is supposedly divine, utilizing a religious path that is devoid of the true Spirit of God. "The concept of a religious path to enlightenment is found in many eastern religions as well as Greek pagan beliefs, and it has become a central theme of western philosophy in the last few hundred years."[14] Paul, writing under the inspiration of the Holy Spirit, knew this was coming:

> But mark this: There will be terrible times in the last days. People will be lovers of themselves, lovers of money, boastful, proud, abusive, disobedient to their parents, ungrateful, unholy, without love, unforgiving, slanderous, without self-control,

---

[10] Henry Morris, *The Long War Against God* (Green Forest, AR: Master Books, 2000), 256.

[11] Andrew M. Sibley, *Cracking the Darwin Code* (Colyton, Devon: Fastnet Publications, 2013), 32.

[12] Ibid.

[13] Although it is possible to have a wrong view of biblical creationism and still be saved, if the core doctrines of Genesis 1-2 are misunderstood it is quite possible to miss the true meaning and importance of the Gospel – and hence miss salvation.

[14] Sibley, *Cracking the Darwin Code*, 4.

brutal, not lovers of the good, treacherous, rash, conceited, lovers
of pleasure rather than lovers of God – having a form of godliness
but denying its power. Have nothing to do with such people.[15]

'Having a form of godliness but denying its power' describes the current 'mysticism' approach to religion that ascribes divinity to nature, denying the one true God of his rightful authority over creation. Mysticism is synonymous with pantheism, which is the belief that 'all is god,' and as we will see pantheism is nothing more than atheism with a religious covering. So, although some people may claim to be 'spiritual,' they are, in fact, nothing more than 'atheists unaware.' Whether in the context of full-blown atheism or mystical pantheism, evolutionism pervades the thinking of many people worldwide. Therefore, as defenders of the faith we must understand the history behind this ancient 'doctrine of demons.'

Admittedly we cannot blame every false teaching (ancient or modern) on Satan. We should not be like the Church Lady from Saturday Night Live, who blames Satan for everything. Fallen man has been successful at devising godless ideas on his own. However, based upon the fact that evolutionism has been successful at drawing people away from the truth of biblical creationism for such a long time, we can safely say that it is a well-utilized 'doctrine of demons.' King Solomon wrote that there are never any new ideas in our time (Ecclesiastes 1:9), and as we will see this is clearly the case when it comes to evolutionism.

Evolutionism & Creationism: Two Chief World Systems

In 1632 Galileo Galilei wrote *Dialogue Concerning the Two Chief World Systems*, which compared the Ptolemaic and Copernican views of our solar system.[16] Likewise, there are two 'chief systems' regarding origins. Evolutionism maintains that time, space, and matter-energy 'exploded' into existence at some time in the very distant past. This was not just millions but billions of years ago.[17] This explosion is commonly referred to as the 'Big Bang' today, and God may or may not have been behind it – it depends upon who you ask. Atheistic evolutionists believe there is no God, so the Big Bang 'just happened.' (It is worth noting that some atheists, as well as pantheists, still believe in a universe which had no beginning and has always existed. The idea of an eternal universe, or even the 'multiverse' idea, is becoming quite popular again. The reason behind this is twofold: First, secular Big Bang models have

---

[15] 2 Timothy 3:1-5, NIV.

[16] The Ptolemaic (geo-centric) system maintains that the earth is the center of our solar system, with the sun and planets revolving around it. On the other hand, the Copernican (helio-centric) system holds that the sun is the center of our solar system, with the earth and other planets revolving around it. Today we know beyond a shadow of a doubt that the Copernican system is scientifically correct.

[17] Both secular scientists and old earth creationists maintain that the 'Big Bang' occurred 13-14 billion years ago.

serious scientific problems, and second is the recognition of theistic implications behind a beginning, which is often too much for secular-minded man to swallow. We will explore both reasons in later chapters.) Theistic evolutionists, on the other hand, confidently maintain that God was behind the Big Bang, as well as every other major step in the creation process. From this explosion, our solar system, planet, and all life on the earth developed over eons of time – once again, with or without God in the picture.

The other 'chief system' is creationism, which maintains that at a certain point in the distant past God created time, space, and matter-energy. Some creationists maintain that the universe is as old as evolutionists say it is – these folks are known as old earth creationists – whereas others are confident that both Scripture and science point to a universe which is only thousands of years old. Although often referred to as 'young earth creationists,' I like the term used by Philip Bell: Historic special creationists.[18] Bell, who serves Creation Ministries International as their CEO for the United Kingdom and Continental Europe, uses this term to distinguish those who maintain that God specially created all life in the recent past from their theistic evolutionary counterparts. I also prefer the term 'recent creationist' as well. (I will use the term 'recent creationist' throughout the remainder of the book, more for the sake of brevity than anything else.) For creationists, only microevolution (small-scale evolution) is a fact while macroevolution (molecules-to-man evolution) is a philosophy of origins devoid of both scriptural and scientific support. God created everything in the universe, including all life on the earth, with intentional design. Only one of these beliefs – evolutionism or creationism – can be correct. As we will see, evolutionism is an ancient philosophy that flies in the face of both Scripture and science.

Although evolutionism and creationism are the two chief systems regarding origins, there are several different viewpoints within each of these positions. Therefore, instead of an 'either-or' situation we end up with a spectrum of views, with some positions more accurately described as evolutionism while others best fit under the banner of creationism. This can make it tough to write on the subject. On one end of the spectrum is naturalistic evolutionism, which claims there is no God in the picture. On the other end of the spectrum is recent creationism, which proclaims that the God of the Bible created everything in six literal, 24-hour days about 6,000-10,000 years ago. For many people, the origins debate ends right there, with these two positions only; the topic of origins is very black-and-white for some folks. But we must keep in mind that there are several other positions as well, all lined up between these two extremes.

---

[18] Philip Bell, *Evolution and the Christian Faith* (Leominster, England: Day One Publications, 2018), 23-24. In his notes on page 24, Bell mentions that some young earth creationists prefer to be labeled 'biblical creationists,' but old earth creationists and theistic evolutionists believe that they, too, are biblical creationists in every sense of the term. Therefore, this can be quite insulting to old earth believers. Two of my mentors are old earth creationists, and they consider themselves to be biblical creationists – as do I. They are fully committed to God in every way.

Theistic evolutionism maintains that God used evolution as his means of creating everything, over billions of years. Like creationists, theistic evolutionists differ in their view. Mostly this involves how often God inserted himself into the creative process. Some theistic evolutionists maintain that God 'snapped his fingers,' so to speak, to initiate the process of creating, and then he distanced himself from further direct involvement in the creation. In other words, after getting the process of evolution jumpstarted God simply sat back and observed how the laws of nature played out. These folks believe that God 'hardwired' into nature everything needed for evolution to 'do its thing.' Some theistic evolutionists insist that, once man had evolved to his current physical state, God breathed into the nostrils of one man the breath of life (Genesis 2:7), thereby directly inserting himself into the process of creation at least one more time. Other theistic evolutionists see God directly involving himself in nature several times, making constant adjustments to the creation as needed. For these folks, the 'natural tendency' of evolution is not that trustworthy, therefore God needed to guide evolution constantly.

There is also pantheistic evolutionism, which is the bridge between the naturalistic and theistic versions. Pantheistic evolutionism proclaims that everything emanates from the impersonal god-force of the universe, which is nothing more than the embodiment of the laws of nature. (Think 'The Force' from *Star Wars*.) However, this view is essentially synonymous with naturalistic evolutionism in that there is no personal God in either view.[19] I am convinced that pantheistic evolutionism exists because of the spiritual void that is inherent in naturalism. This lack of spiritually in naturalism (materialism or atheism) is too hopeless and depressing for many people, who need something bigger than themselves to hold onto.

Old earth creationism teaches that the God of the Bible specially created everything over billions of years, but without resorting to large-scale evolution. Within old earth creationism, there are a few major views. Day-age creationism teaches that each of the six days of Genesis 1 were long periods of time, while progressive creationism teaches that the six days were indefinite time periods with overlap between the days. Punctuated 24-hour creationism insists that each of the six days were 24 hours long, but between each of the days was enough time to accommodate the vast ages needed for an old earth.

Gap theorists combine an old earth with a straightforward reading of the Genesis text. Proponents of the gap theory insist that between Genesis 1:1 and 1:2, or possibly even between Genesis 1:2 and 1:3, are all the long ages needed for cosmic and geologic evolutionism, as well as the development of the earliest life forms on the earth. Instead of rendering Genesis 1:2 as, "Now the earth was

---

[19] The term 'personal God' implies that God possesses the attributes of personality, such as emotion, rationality, volitional will, and so forth. It also means that God is able and willing to have a personal relationship with his creatures. Personality is not something that we would expect from an impersonal god-force.

formless and empty," gap theorists confidently assert that the verse should read, "Now the earth became formless and empty." Some great cosmic catastrophe, usually associated with the fall of Satan and the ensuing 'Lucifer's Flood,' destroyed the primeval earth.[20] After long ages had passed, in which the earth laid in ruins, God re-created everything in the recent past as outlined in a straightforward reading of Genesis 1. Essentially, this view combines either theistic evolutionism or one of the old earth creationist views with recent creationism. This explains why some adherents of the gap theory instead prefer the term 'ruin-reconstructionism' instead. Besides these numerous views, there are other positions as well, all modifications of the views described.

Therefore, the spectrum might look something like this: Naturalistic evolutionism, pantheistic evolutionism, theistic evolutionism, progressive creationism, day-age creationism, punctuated 24-hour creationism, gap theory, recent creationism.[21] If the average Christian, who often thinks in terms of either naturalistic evolutionism or recent creationism, heard someone say that they believe in the gap theory, but with theistic evolutionism explaining the original (pre-catastrophe) earth, he or she would probably have no idea what is meant by that statement. Hence, one of the reasons for this section of the book.

---

[20] Of course, there is no 'Lucifer's Flood' mentioned in Scripture. Rather, gap theorists insist that a devastating flood, far greater in scope than Noah's Flood, could only be attributed to Satan's fall.

[21] Some people might quibble over where to place these views, maybe using a slightly different order, but by-and-large this is a correct arrangement.

Table 1.1

Spectrum of Views on Origins

Naturalistic Evolutionism
↓
Pantheistic Evolutionism
↓
Theistic Evolutionism
↓
Progressive Creationism
↓
Day-Age Creationism
↓
Punctuated 24-Hour Creationism
↓
Gap Theory
↓
Recent Creationism

The battle between evolutionism and creationism has seen many players throughout history, some serving God while others serve Satan – many inadvertently and maybe even a few knowingly. Moses, writing under the inspiration of God's Spirit, revealed creationism to us in the book of Genesis, while the Egyptian priests of his time promoted evolutionary ideas. In the first century BC the Stoic philosopher Cicero argued for creationism while Lucretius and the other Epicurean philosophers of the time promoted evolutionism in their teachings.[22] In the second century AD the Christian apologist Theophilus battled the skeptic Galen over the issue of origins, and the battle raged on throughout the centuries. Thomas Henry Huxley, better known as 'Darwin's Bulldog,' and Samuel Wilberforce, the Bishop of Oxford, formally debated each other only one year after Darwin's *Origin of Species* was published, and the debates – formal, informal, and now online – continue to this very day. The battle for origins has always been fierce, and it shows no sign of letting up anytime soon.

My Story

Before going any further in our study, let me first tell you a little bit about my background. Whenever I read a book, I always like to first know a thing or two

---

[22] Keeping in mind that Cicero, as a Roman and not a Jew, was an 'Intelligent Design' proponent rather than a biblical creationist. He was a Stoic philosopher who saw the evidence for special creation.

about the author: What makes him or her 'tick,' and how does the author's background reflect upon the content of the book. In the case of this study, you might find it interesting to know that I was once an evolutionist who was very skeptical of religion in general, and Christianity in particular.

I grew up in Galesburg, Illinois, a small city half-way between Chicago and Saint Louis and less than an hour from the Illinois-Iowa border. I was raised in a good family which taught values and respect for others, although I may not have always been good about putting that training into practice. Other than being baptized as an infant, church was not a routine part of my early life. By the time I entered junior high school I avoided church because I felt uncomfortable there, mostly due to my unfamiliarity with Christian doctrine and practice. By high school I had adopted the full-blown skeptical position that organized religion was for weak-minded people who needed someone to tell them what to think. I came to this conclusion after hearing the skeptical claims of some very influential acquaintances. I labeled myself a freethinker, as many atheists do, but unlike atheists I always believed in God because I could not accept the idea that the universe began by chance. I did, however, believe that molecules-to-man evolutionism was God's means of creating the diversity of life found on the earth, both now and in the fossil record. I believed evolutionism to be an indisputable fact, truly beyond question. What sort of weirdo would doubt evolutionism, anyway?

I was sure that religious beliefs are of no importance to God, as God must only be concerned with a person's moral values and treatment of others. The so-called 'Golden Rule' was the only religious rule I needed back then. I viewed the Ten Commandments as nothing more than outdated Hebrew ethics, although some of the commands were noble and worthwhile. I believed that Jesus was a great moral teacher, and I loved how Jesus stood up for the poor and oppressed, but I did not believe that Jesus is God. I believed that the Bible is a holy book, but no more holy or truthful than the Qur'an, the Hindu and Buddhist writings, or the Book of Mormon. Ultimately, all these religious books were purely man-made in my mind. I believed that God would accept almost everyone into his presence upon death, excepting a few evil misfits like Adolph Hitler and Joseph Stalin.

As you can see, none of my beliefs were biblical. I was pretty much the 'poster child' for secular humanism back then. I was theologically confused simply because I never took the time and made the effort to examine the evidence for the biblical worldview.

But then everything began to change. At twenty-three years of age, I was accepted into a Radiation Therapy Technology residency program in Rockford, Illinois. The program lasted for one year, from September 1987 to September 1988, and it was the most intense year of my life. This was the year that began my spiritual journey from skepticism to Christ. Some interesting things happened that year. First off, my biological sciences instructor turned out to be a devout

Christian creationist. The usual scenario in college involves a devoutly faithful student and a skeptical professor, but I never seem to experience life in the usual way! I admired my biology professor's seemingly endless knowledge of anatomy and physiology. Several times that year she made the point that the human body is way too complex to have evolved slowly in stages, and instead appeared to have been created instantaneously. That got my attention, to say the least. As a result of my respect for her, I found myself questioning my evolutionary beliefs, and I began to think that if I could be wrong about origins, then maybe I was wrong about some other faith-related issues as well – namely the nature of God, and our relationship to God through Christ.

During this time, I encountered some people who challenged me to explore the truthfulness of Christianity. One man told me that if I became a Christian, I would experience an emotional high unlike anything I had ever experienced before. I had no reason to doubt him – he was, after all, extremely sincere about it – but I wanted nothing to do with a religion that promised only an emotional bliss. Especially since my life was going well, anyway. I needed something more than just an emotional high: What I really needed was truth. That was my motivation.

People come to Christ for a variety of reasons. The overwhelming desire to avoid Hell seems to be near the top of the list of reasons for most people. That never worked for me, though. For whatever reason – maybe it was sheer ignorance, or maybe even arrogance – Hell-avoidance was not much of an incentive for me. Then there is the flip side of that reason: Heaven-attainment. That seems a bit more positive to me. However, as an extremely healthy weightlifter in his late twenties, Heaven seemed like something far away – and not to be concerned about at the time. (Heaven could wait, as far as I was concerned.) Like Pontius Pilate two millennia earlier, what I really wanted to know was 'truth.' Yet unlike Pilate, I asked the question, "What is truth?" with nothing but sincerity in mind. I was on a pursuit for the truth about life and its big questions. I knew my former secular ideas were on shaky ground, but that did not mean that Christianity was automatically true by default. I had to test it first, and there would be challenges – like trying to reconcile a good and loving God with so much suffering, death, and evil in the world. That was a tough one.

I always knew that suffering existed in the world, and obviously I knew that people died, but during this year I was constantly confronted with the problem of suffering and death. Suddenly it became very real to me, more than just intellectual 'head knowledge.' I was warned that the program would be emotionally challenging, and it was. Although I hated to see anyone battle cancer, it was the children whom I treated that melted my heart – and caused me to have many intense conversations with God. During some of those conversations I was quite angry. At this point I could have gone the route of atheism, as many do because they cannot reconcile the problem of suffering and death with the loving God who is revealed in Christ. But I always knew God existed because of the intelligent design that is seen throughout all of nature. This was reinforced through my studies in the natural sciences during this year. I concluded that, regarding the problem of suffering and death, God knew something that I did not. This must be the greatest understatement in history!

After exploring the biblical worldview for a few years, I came to a point where I knew the claims of the Bible were true. This had not been easy, however. I struggled with two issues in particular: The problem of suffering and evil, and salvation in Christ alone. How can I – or anyone, for that matter – reconcile the biblical God of love with so much suffering and evil in the world? This was a huge stumbling block for me, but an even tougher issue was salvation in Christ alone. What about all those Hindu's, Buddhist's, Muslims, and even secularists who are good people, but not professing Christians? How could God allow them to be separated from him for eternity, their only crime being unbelief in the biblical God? (Plus, many of them never even had access to the Bible during their lifetime.) Over time I discovered that I was not alone in my struggle: The problem of suffering and evil, and the exclusivity of Christianity, seem to be two of the major skeptical objections to the faith. Over a quarter of a century after becoming a believer, I have never forgotten this time of intense wrestling with the Christian faith – which is why I address these issues in depth in chapters seven and eight.

It became clear to me, however, that the evidence for Christianity was overwhelming. I may not have had all the answers, but I knew enough to become convinced that Christianity was true. I had always said that in everything I examined I would try to put rational evidence before a personal agenda, and I meant it. Finally, at the tail-end of my twenties, I accepted Christ as my Lord and Savior. In some ways, I was just like C.S. Lewis decades before me: On one hand, I felt like I entered the faith 'kicking and screaming' because of the intellectual objections that I had to fight through. (Thank God for the excellent work of the apologists who came before me, who God used to prepare me for acceptance of the faith.) On the other hand, however, I experienced a 'peace of heart and mind' that could only come from God himself. Maybe a lot of apologists have felt that strange dualism as well during their time of conversion. Although I was now a believer, my exploration of the faith was far from over. In fact, it had barely just begun.

Before going any further in my testimony, I must point out something of great importance: I did not come to Christ on my own accord. It may have felt like I initiated this quest for truth, but that is simply not true. Before I ever began to seek after God, God was pulling me in his direction. This is a belief known as prevenient grace, which states that before anyone can come to God, God must first open that person's heart and mind to the message of the Gospel. We do have free will, but nonetheless God draws us to him well before we begin to seek after him. Perhaps John 6:44 reveals this doctrine better than any other verse: "No one can come to me unless the Father who sent me draws them…"[23] There are many other verses which demonstrate the truth of prevenient grace (Luke 24:45; John 12:32; Acts 16:14; Romans 5:8).

For the first year of my Christian life, I remained a theistic evolutionist. I had no reason to doubt that God used evolutionary processes as his means of creating, as the 'Case for Christ' was the focus of my Christian exploration up to this point. It was only later that I would explore the 'Case for a Creator' in detail.

A year into my Christian walk, my wife brought home a book for me to read. The book came from one of her co-workers, Anita Reese, who had thought I might enjoy it. The book was Henry Morris' *The Biblical Basis for Modern Science*. I had never read a book quite like this one: It challenged me to the core of my being. I stayed up well into the night reading it, despite knowing that I would be exhausted at work the next day. (And I was!) But I could not put the book down. Everything Morris wrote made sense to me, although I must admit that the idea of a young earth and universe was a radical idea for me at the time. I had never heard of such a thing before reading this book.

A month later, I saw Dr. Morris' son John – also a very accomplished creation speaker – at a church in Bloomington, Illinois, which is not far from my hometown. The younger Dr. Morris, along with the highly engaging Frank Sherwin, delivered several presentations on the evidence for recent creationism. I was bitten by the creation science bug, and the affliction has not been cured to this day!

Since becoming a Christian creationist, I reject molecules-to-man evolutionism in every way, shape, or form. It does not make sense to me, scripturally or scientifically. I now see it for what it is: A secular philosophy, and a humanist worldview. The only area where I have struggled over the years has been the age of the creation. When I first became a Christian, I aligned with the young earth camp. They were, after all, the only version of creationism that I even knew about at that point in my faith journey. However, I eventually learned about another creationist position that challenged my thinking in this area. Old earth creationism became a new idea for me a short time after becoming a creationist. The most successful old earth creationist ministry was, and still is, Reasons to Believe. This ministry was founded by the Canadian astrophysicist Hugh Ross. At first, I was not all that excited about exploring old earth creationism, as the young earth view made so much theological sense to me that I felt no need to explore any other position – at least at that point in my investigation of the faith. However, a young earth creationist inadvertently paved the way for my interest in old earth thinking.

This young earth creationist, who will remain anonymous, wrote that no one should buy any of Hugh Ross' resources. Big mistake. For me, this was akin to a teacher or principal telling a young student that a certain book is on the 'banned list.' Never mind that the student may have had no desire to read the book before

---

[23] John 6:44, NIV.

hearing that. Now, upon learning that the book is banned, the student must read the book, at all costs! My reasoning went like this: There must be something that young earth creationists do not want the Christian body to know about old earth creationism. Before long, I had read through several of Dr. Ross' books, and I found him to be a devoted believer with a heart for spreading the message of creation and the Gospel. I still believe that. The man is my Christian brother, whether I agree with him or not. I found that, concerning cosmology, I appreciate Dr. Ross' expertise. However, anthropology was another matter: I disagree with the old earth position on mankind's origins, although I believe that it is possible that there could be gaps in the biblical genealogies that extend mankind's origin a few thousand years at best.[24] Even some of the young earth material from a generation ago held to a 6,000-10,000 year-range for the creation, so that is not exactly a heretical idea. However, most recent creationists today will not consider any number much beyond 6,000 years for the creation, as they typically hold fast to the genealogic dates given in the Masoretic text. I disagree with Dr. Ross' position on Neandertal man being less than fully human and an origin for Adam as far back as 100,000 years or more, but I do appreciate his enthusiasm for spreading Christ's message to the world. We should all acknowledge his Christian commitment, whether we agree with his ideas or not.

During my time engaged in seminary coursework and the apologetics certificate program, I wrote some of my papers from an old earth perspective. However, I routinely read young earth resources and would attend presentations given by recent creationists. Scientifically, I believed that the earth and universe could be quite old, but scripturally I leaned toward a young creation. I investigated the issue to the best of my ability, but it was a confusing time. I would read something convincing from a recent creationist, and then I would wonder why I was not totally devoted to that position. But then a few days later I would read something equally convincing from an old earth creationist, and then I would remember why I was straddling the fence. I eventually declared myself a 'general creationist,' neither young earth or old earth, just taking it all in and trying to figure it out. Someone would say to me, "But 'so-and-so' is a young earth creationist, and he is a brilliant scientist." That sounded good, until I remembered a second later that some other equally incredible scientist was an old earth creationist. There are amazing scientists on both sides of the fence. It is better to go with the theologians instead, I reasoned. But guess what? Great theologians and apologists hold to both positions as well, and they are all using Scripture as their guide. Maybe I need to switch to an easier topic, like predestination versus free will!

Around this time, I was speaking with a man who had been a founding member of Jews for Jesus. He came to our church at least once a year, conducting

---

[24] The claim that there are gaps in the biblical genealogies is common among old earth creationists, and as we will see in chapter six this idea has merit. However, I believe that these gaps extend mankind's history by only a few thousand years at best, and likely less than that.

study conferences that were nothing short of amazing. We financially supported him as a missionary, and we were blessed through his teaching ministry. I asked him his opinion on the age debate. As a Jewish scholar with a background in Hebrew, theology, and even a lot of philosophy coursework in college, I knew his ideas would be well informed. He told me that, regarding origins, it could go either way: The creation could be young, or it could be old. In terms of the Hebrew language, there was no way to say for sure, at least in his mind. I have no reason to doubt the truthfulness of what he said, for every Hebrew scholar who says the creation is young, there is another who says it is old. Maybe I was looking for a more black-and-white, concrete answer, but I appreciated his knowledge and his honesty. Sadly, he passed at a young age. I, and everyone in my congregation, misses him greatly.

When I wrote my first book, I approached the topic of origins from my 'general creationist' position. In *Worldviews in Collision*, I noted the following: "I personally utilize resources from both young earth and old earth creationist ministries, and rather than focusing on the age of the earth I instead emphasize the evidence for a beginning of the universe, the incredible design in nature, and the fact that life comes only from life."[25] That is still my approach when dialoguing with unbelievers. I have no reason to doubt that the universe is young – we will get to that in a moment – but the age debate is not the focus of my ministry. There are other people who do that very well, far better than I do. I see myself as being more of an intelligent design proponent who personally prefers a recent creation for hermeneutical reasons.[26] When sharing my faith with an unbeliever, I will always stress the fact that the universe had a beginning, over the evidence for a young creation. If the age issue comes up, I will go there in a heartbeat. There is, after all, a place for that topic, too – and the evidence for a recent creation does sometimes convince unbelievers that evolutionism is not true. In apologetics, everything has its place.

After the publication of *Worldviews in Collision*, I decided to give recent creationism another look. I have no idea why I felt the need to do that, but the thought was overwhelming. I will not go so far as to say that my re-examination of recent creationism was Spirit-inspired, but I felt compelled to re-examine the issue with an enthusiasm that I had not possessed for quite some time. Even from the time that I began to explore old earth creationism several years ago, I always kept one foot in the recent creationist camp. Therefore, it was not like I was coming back to a doctrine that I had previously disavowed. As I said before, I spent a long time straddling the fence between the young earth and old earth positions. I was, in essence, a 'man without a country,' not fully accepted by either side. That, by the way, is not an easy position to be in.

In my church library, there were two books that caught my attention: *Old-*

[25] Randall Hroziencik, *Worldviews in Collision: The Reasons for One Man's Journey from Skepticism to Christ* (Greenwood, WI: Paley, Whately, and Greenleaf Press, 2018), 58.
[26] Hermeneutics is the branch of theology concerned with the literary interpretation of the Bible.

Earth Creationism on Trial, by Tim Chaffey and Jason Lisle, and *The Genesis Factor*, edited by Ron Bigalke, Jr. I read through these books, and my confidence in the young earth position blossomed. Shortly after that, I enrolled in Answers in Genesis' online courses, and my confidence in a recent creation was further strengthened.

Shortly after my re-dedication to recent creationism, I spoke to one of the leaders in a major creation science organization regarding the possibility of becoming an adjunct speaker. Since *Worldviews in Collision* was already on the market, I explained to him that I have written on creationism in the past from both an old earth as well as a general creationist position.[27] The conversation immediately went downhill from there. He focused on the fact that I had been a recent creationist, then switched to the old earth position, then became a general creationist, and now claimed to be back in the young earth camp. Basically, he did not trust me. Not that I blame him, mind you. I do see his point. I was honest, however, and explained my story to him. The possibility of doing any adjunct speaking for that group was dead and buried. To be honest, my chances of joining any young earth ministry are slim-to-anorexic. No one wants the guy who seems a little too friendly to those who believe differently about origins.

A lot of people, especially those who are skeptics, will quickly dismiss me as being nothing more than a religious nut with no real idea of how science works. I beg to differ, of course. I feel that my educational background, combined with my passion for creationism and apologetics, has prepared me to tackle the issue of origins quite well. My undergraduate degrees focused on the natural sciences, while my graduate work was in theology and apologetics. This does not mean that I am the world's premiere expert concerning origins. Not even close. It does mean, however, that I have not only studied the topic on my own in great depth, but much of my formal education has helped me tremendously in this area of study. I feel that I have a voice in this matter. (As we all do.)

So, just what is my position on science and origins today? I believe that science should be divided into two broad areas: Operational science and historical science. Operational science is concerned with observation and repeatable experimentation. Anyone can perform operational science, regardless of their religious persuasion or philosophical worldview. The atheist, the Christian, the Hindu, and the Mormon should all be able to record scientific observations and conduct experiments with the same results. Operational science is not really part of the origins debate, however. Historical science, on the other hand, is focused on how things originated. Unfortunately, there are no repeatable experiments that can be performed which will shed indisputable light on this. Concerning the origin of the universe, one scientist may be convinced that an explosion from nothing – and for no reason – brought everything into existence over billions of years, while another scientist may conclude that, "In the

---

[27] Besides *Worldviews in Collision*, I also have papers posted on the Evidence for Christianity website (evidenceforchristianity.org).

beginning God created the heavens and the earth"[28] not all that long ago.

I believe that a beginning for the universe, the fact that life comes only from life, and the complex information found in the genetic code proves creationism. I also believe in biological evolution on a small-scale (microevolution). Living things do change over time. Natural selection is a fact, but it only works to keep a species strong, not to change one type of plant or animal into something entirely different. But the philosophy of molecules-to-man evolutionism is nothing more than a secular worldview with no scriptural or scientific support. It is 'science falsely so called' (1 Timothy 6:20). To deny God's creative power and design is to deny the obvious (Romans 1:20).

Throughout this book I do discuss the age of the earth and universe. Christians love to fight about it, and unbelievers – as well as some believers – love to make fun of recent creationists. As already mentioned, the age debate is not the focus of my apologetics ministry. I am convinced that the beginning of the universe is of far greater impact than the age debate, therefore that is the focus of my ministry. Personally, I like the recent creation view. It makes scriptural sense to me, and it makes sense of the big picture theologically. But I have always struggled with only 6,000 years for the age of the earth. I have always believed that the period between the Flood and Abraham, at least according to the genealogies given in the Masoretic text, is simply too tight to allow for everything that needed to happen in that time. The Septuagint, the Samaritan Pentateuch, and the Jewish-Roman historian Flavius Josephus provide numbers that make more sense to me. I still believe in a young earth, but not necessarily 6,000 years young. I will discuss this in more detail in chapter six.

Today, I consider myself a creationist, period. Although I prefer the recent creation view, I will listen to the ideas of old earth creationists, gap theorists, and even theistic evolutionists. (I am more open to old earth ideas, but I am less unconvinced by the beliefs of gap theorists and especially those of theistic evolutionists.) The different views on creation have a few broad points in common that deserve to be recognized – namely a universe with a beginning and a God who oversees his creation – but our differences are significant. We will explore this, and other issues in more detail as we proceed through our study. But for now, let me conclude this section by noting one thing of utmost importance: I will not denigrate believers who hold to an opposing view on creation. As believers, we must learn to become kinder, gentler, and more respectful apologists while we make the case for our view on origins. It has been said that people do not care about what you know, until they first know that you care. I agree with this sentiment. By showing respect for others, whether they are believers or not, we can better make the case for creationism. As the Apostle Peter wrote, state your case "with gentleness and respect."[29] I cannot do that by referring to others as 'compromisers,' a term which I find repugnant. My mentors

---

[28] Genesis 1:1, NIV.
[29] 1 Peter 3:15, NIV.

are old earth creationists, and they are anything but compromisers when it comes to God's Word. They are world-class theologians 'who correctly handle the word of truth' (2 Timothy 2:15) better than most people I know. As already mentioned, we will explore the topic of the age of the creation in more detail in chapter six. You may want to skip ahead to that chapter at this time, before coming back here.

What We Will Explore

As I went through this book one last time before publication – mostly to check for spelling and grammatical errors – I realized that some people might conclude that this is really two books in one. This was never my intention, however, as each chapter was written with an overarching purpose in mind. The first six chapters focus on the origins debate, while the last three chapters have more to do with theology than science. This is by design, however: It has been my experience that most skeptics are not skeptical mainly because of science. Rather, it is the deeper issues of theology that seem to be the real problem, namely suffering and evil as well as salvation in Christ alone. As a former skeptic, I get that. Although not every evolutionist is a skeptic of Christianity, I wanted to address these theological objections to the faith in some detail, for the benefit of those who are skeptical but nonetheless open to reading material from a Christian writer. More likely, this book will find its way into the hands of faithful believers who know someone who is skeptical, and they are looking for a resource that will help them in their continuing interactions. Either way, I hope this book is a blessing for all who read it.

In the first chapter we will examine the rise of evolutionism in the ancient world. In its earliest form that we know about, evolutionism began with the great civilizations of Mesopotamia, Egypt, and the Indian sub-continent, and later became more formalized in Greece and Rome. It may be that evolutionism preceded even these great civilizations, perhaps going all the way back into the pre-Flood world. Only God knows if this is the case, but a little speculation is not necessarily a bad thing.

In the second chapter we will examine how the early Church confronted evolutionism. Even by this time in history, the battle between evolutionism and creationism was long-standing. The early Church leaders and apologists knew the importance of the battle for origins, as evolutionism could potentially weaken the witness of the Christian faith. (Spoiler alert: It has done that very thing.) Chapter three will discuss how the Church confidently proclaimed creationism in the Middle Ages, but then faced great opposition to the veracity of Genesis in the Renaissance and Enlightenment eras. This is when the slide into our current cultural skepticism began to take hold.

In the fourth chapter we will examine evolutionary thinking from the time of Charles Darwin to the present. After thousands of years, the battle between naturalistic and biblical thinking on the topic of origins has only intensified. Chapter five is focused on building the 'Case for the Creator,' while the next four chapters focus on addressing the objections of skeptics. Chapter six – the longest chapter by far, but in some ways the most interesting – recounts the age of the earth and the debate over death before Adam, among other things. In chapter seven we will wrestle with the problem of suffering and evil. As already mentioned, it has been my experience that evolutionism is not usually the main reason for most people rejecting the God of the Bible. But what is the number one reason? Although skeptical opinions vary, of course, it is often the claim that pain, suffering, and evil proves there is no God. This is an important topic that cannot be avoided, and we will all do well to wrestle with it.

In chapter eight we will examine the especially difficult topic of divine justice, namely the eternal fate of the unevangelized as well as the reality and nature of Hell. Some books, such as Norman Geisler's *If God, Why Evil?* and Clay Jones' *Why Does God Allow Evil?* include the fate of the unevangelized within the umbrella of suffering and evil, but I am assigning it a chapter of its own because it is such a huge concern for skeptics and believers alike. Finally, in chapter nine we will focus on the issue of our mortality, and how it fits into the big picture of life with and without God. We will then conclude with a summary of the major points offered in this 'big picture' study.

At the back of the book, I have supplied a list of resources for further study on the origins debate. As I mentioned in the acknowledgements section, this book is a good introduction to the topic of origins, but it does not address every issue involved. Not everyone will want to explore evolutionism versus creationism beyond what is offered in this book, but for those who do I want to provide some trusted resources for continued study.

After the list of resources for further study, I have supplied a glossary of the important terms that you will encounter in this book. Some readers may be extremely well-read regarding the origins debate, while others may be relatively new to the topic. Either way, it is not a bad idea to skip ahead to the glossary of terms at this early point in the study. For those who are new to this topic, you can avoid the frustration of encountering a term that is unfamiliar to you by skipping ahead now and reading through the glossary. On the other hand, if you are an advanced student of the origins debate you may find that my definition of certain terms is not 100% in agreement with yours. Although I suspect that our definitions will be quite close, it may be worth skipping ahead now to the glossary and determining this for yourself.

It is my sincere hope that all who read this book will not just learn specific things of interest, but more importantly will begin to see the 'big picture' involving the spiritual battle that plays out behind the scenes. The history of the origins debate has unfolded throughout the entirety of human history, but I do believe there is another, even deeper and more sinister story going on in the background. We must consider this 'other level' of the origins debate as well. Enjoy.

CHAPTER ONE

Evolutionism in the Ancient World

As already mentioned, evolutionism goes back at least to the time of the ancient Greeks, and in its earliest forms even to the time of the Babylonians, Egyptians, and Hindu Brahmins. But was this philosophy present even before these great civilizations arose? Although admittedly speculative in parts, we should consider the possibility of evolutionism in the pre-Flood and early post-Flood world – from the time of Adam and Eve to the Tower of Babel.

Evolutionism is one of the reasons given by atheists and agnostics today to justify their skepticism. In fact, when it comes to reasons for denying God, evolutionism has been a major excuse for a long time. So, when considering the history of evolutionism why should we stop at the Greeks or even the Babylonians? History does precede even these great civilizations.

We will explore the history behind the evolutionism-creationism debate in not only this chapter, but the next three chapters as well. It has always been my conviction that if one wants to really learn about a topic, he or she must begin with the history behind that topic. Only by understanding the history involved in the origins debate, will the evidence for God and creation in chapters five and six begin to make sense.

Evolutionism in Genesis 1-11

There was a time in history of which we know little about today, save for what God has revealed to us in the Mosaic writings. It is true that secular-minded archaeologists and historians occasionally discover 'hints' of this time, but by and large this period of history remains a mystery for much of mankind.

The first book of Moses begins with God's revelation of how the universe, this world, life and especially humanity came into existence. It was clearly through an act of intentional creation, with each plant and animal type being created 'after its kind' (Genesis 1:11-12, 21, 24-25). God created mankind 'in his own image' (Genesis 1:27), hardwiring us to know that we will live for eternity (Ecclesiastes 3:11). Fourteen hundred years later, Paul confirmed these truths when he wrote that all people know of God's existence because we can plainly see God through his works, being so obvious to us that no one can be excused for failing to acknowledge the Creator (Romans 1:20). Additionally, our inherent 'moral compass' or inborn ability to know right from wrong speaks to God's creating us in his likeness (Romans 2:14-15). The atheist truly has no excuse for his disbelief in God.

The first two chapters of Genesis reveal the plan behind God's handiwork. These are amazing chapters, filled with the hope of a new creation. Yet all was not well for long. We barely get through these chapters and everything heads south. Satan, the 'Father of Lies' (John 8:44), deceives the first created people. Man becomes a fallen creature, inherently corrupted from this time forward

(Genesis 3:16-19; Romans 3:23). In fact, mankind became so corrupt, and so wantonly separated from God, that the Lord was grieved that he created humanity in the first place (Genesis 6:5-6). How can a race of people, made in God's image and hard-wired to know the truth about their Creator, become so desperately wicked in such a relatively short time that God destroys almost all of them (Genesis 7:1)? The explanation is surely multi-faceted, but what if the 'Great Deceiver' was able to somehow convince people that they were not really created by God in the first place? If he could do that, people would not feel the need to offer worship, and thereby give allegiance to, a creator-god of any sort. Instead, people would do what they pleased without any thought of divine accountability.

At this point in our study, we need to pause to examine the person of Satan once again. (We did this only briefly in the introduction.) The unbelieving world laughs at the idea of Satan being real, which should come as no surprise since many of these same folk's scoff at the idea of God's existence as well. I remember shortly after 9/11 watching a segment on Fox News about the reality of Satan. It is sad that it took 9/11 to strike up this conversation, but it did. The 'problem of evil' was on the hearts and minds of people everywhere, and not just in America. People wondered if evil actions were merely the result of inherently bad behavior in people – perhaps even our 'left-over animal ancestry' rearing its ugly head – or if there really could be a spiritual entity behind it all. As a society we talked about it for a while, but then like most things of a spiritual nature we tend to forget about it eventually. The idea that Satan is not a real being, but rather a mythical character intended to explain evil in the world is no doubt an idea inspired by Satan himself; it is, after all, impossible to defend against an enemy that one does not believe in. The Fox News segment reported positively on Satan's existence. (That is, positive from my Christian perspective.) Even for many people who doubt or oppose the spiritual realm, Satan was real – at least for a while.

Scripture repeatedly notes the reality, and character, of Satan. It is never flattering. Seven books in the Old Testament teach the reality of Satan (Genesis, 1 Chronicles, Job, Psalms, Isaiah, Ezekiel, Zechariah), 19 of the 27 books in the New Testament refer to Satan as a reality – which includes all of the known New Testament writers – and Christ himself refers to Satan some twenty-five times. Thousands of Satanists throughout history have attested to the fact that Satan is real, and doubtless some of them encountered him directly. Satan's influence permeates this world. Although Satan is a created being and therefore has limitations – he is not omnipotent, omniscient, or omnipresent as is God – he nonetheless has two incredible powers that man does not possess. First, having existed throughout the history of the world Satan knows from experience how people will most likely react in any given situation, based upon his countless observations of people. Second, Satan is believed to rule over one-third of the angelic realm (Revelation 12:4), therefore he is the master of a vast network of malevolent beings who do his bidding. Fortunately for believers, Satan and the fallen angels cannot defend against God.

Satan is our adversary in every way (1 Peter 5:8). In fact, the name Satan literally means 'adversary.' Satan is also called the 'Devil' (Matthew 4:1). The word 'Devil' carries the idea of both adversary and slanderer, as Satan not only slanders man to God, but slanders God to man (Genesis 3:1-5). Satan is our enemy (Matthew 13:37-39), the 'father of lies' (John 8:44), and a murderer (John 8:44). The word 'murderer' literally means 'man killer.' Worse than physical death, Satan can kill the soul through leading a person astray. Satan is called the 'god of this age' by Paul (2 Corinthians 4:4). This does not mean that Satan is deity, but rather in this evil age Satan is its god in the sense that he is the earthly head of fallen mankind. Satan is behind the false religions and philosophies that have plagued mankind for millennia. Sadly, Satan possesses the ability to blind the minds of unbelievers, preventing them from seeing the truth found in Jesus Christ (2 Corinthians 4:4). As we have seen, Satan possesses great power. However, that power is limited and temporary. In the end, Satan cannot win in a war against God.

Satan hates God, but he cannot do anything to harm God directly. Therefore, he strikes out against God's creation, especially the pinnacle of God's creation – mankind. What better way to lead people astray than to convince them that there is no God, and that we are nothing more than a cosmic accident. He may have even convinced himself that he evolved from the 'primordial waters' described in Genesis 1:2 and some of the ancient pagan cosmologies.[30] Satan might find it more comforting to believe that he evolved sometime in great antiquity, rather than being created by the one whom he opposes. Just maybe he deceived himself in this way. Interestingly, the ancient Greek poet Hesiod wrote *Theogeny* ('generation of the gods') to explain how the pagan gods arose through self-generation. Did Hesiod come up with that idea on his own, or did he get that idea from someone else? Perhaps Satan and the fallen angels really believed that they

---

[30] Examples of ancient pagan cosmologies that also refer to the primordial waters, or 'the deep' of Genesis 1:2, are the Egyptians, the *Enuma Elish* of Mesopotamia, and Thales of Miletus. The Egyptians taught that before the beginning of creation there existed an infinite sea of water that was dark and chaotic. There was nothing above or below this sea, as the sea was all there was. Then Atum, the creator-god of the Egyptians, brought himself into existence by separating himself from these waters. An interesting point to keep in mind is how different this is from the Genesis account of origins, where the one true God eternally existed before the creation. However, this idea of Atum bringing himself into existence from the waters of 'the deep' fits nicely with Hesiod's *Theogeny*, which is explained in the text above. In the ancient Mesopotamian *Enuma Elish*, creation begins when the pre-existent primordial waters Tiamat (salt water, 'the deep') and Apsu (fresh water) conjoined, producing the first generation of the gods. Once again, this dovetails perfectly with Hesiod's *Theogeny*. Finally, Thales of Miletus, considered to be the first Greek philosopher, taught that everything ultimately came from one source: Water. This would include the gods as well, which is once again a perfect fit with Hesiod's *Theogeny*. This idea that the ancient gods (fallen angels) evolved from the primordial waters of 'the deep' may very well have an ulterior motive behind it.

– the so-called 'gods' (1 Corinthians 10:20; 2 Corinthians 4:4) – were self-generated rather than created, and they in turn imparted this belief to fallen man. Hence Hesiod's *Theogeny*. After all, if these rebellious angels believed they were self-generated, they would have no God above them, to rule over them. Food for thought.

Can we say for sure that evolutionism captivated the warped hearts and minds of people in the pre-Flood world? No, but it does make sense. If evolutionism works well now, and has worked well all the way back to the time of the ancient Greeks and Babylonians, then why not think that Satan used this idea in the world before the Flood as well?

At the time of the temptation in the Garden of Eden, Satan questioned Eve with these words: "Did God really say, 'You must not eat from any tree in the garden'?"[31] It is the first four words that I want to focus on at this time, however. I strongly suspect that Satan began many questions with these four words, and maybe one of the questions was this: "Did God really say that he created you?" Both Adam and Eve, as created beings, would have experienced the moment when they first became 'self-aware.' For Adam, I believe that moment was when God breathed into his nostrils the breath of life (Genesis 2:7). At that initial moment of self-awareness, Adam knew he existed, but how much else did he immediately know concerning his origins? Maybe more than we realize, but maybe God had to fill in the details to a considerable extent. Satan may have capitalized upon this by telling Adam that he was evolved, and that he became self-aware only at a certain point in his evolution. He may have done the same with Eve. Whether true or not, we can say with certainty that Satan has been trying to persuade mankind of evolutionary origins for a long time – likely far longer than we realize.

At the end of Genesis chapter four, we read that at the time of Seth's son Enosh "people began to call upon the name of the Lord."[32] Almost all commentaries note that it was at this time in early history that people began to seek the Lord through prayer and communal worship, as the perfectly holy God and fallen man were no longer in direct communion with each other as they were in the days of Adam and Eve, and even in the days of Cain and Abel. Yet we also read in Genesis that mankind had become so desperately wicked that God was grieved in his heart that he had created man in the first place (Genesis 6:6). Although most people on the earth were wicked and wanted nothing to do with their Creator, there was still a remnant of God-honoring people who acknowledged the Lord. Perhaps these two groups could be partially defined by their view on origins. If so, most people upon the face of the earth were almost certainly evolutionists – or at least strongly influenced by evolutionism in some way or another – while a small remnant of creationists co-existed among them. If that was the case, it sounds a lot like today, doesn't it?

---

[31] Genesis 3:1, NIV.
[32] Genesis 4:26, ESV.

Is it also possible that evolutionism was a religious idea during the time of Nimrod's reign? Since Nimrod rebelled against God and led the people astray, that notion makes sense as well. Ernest Abel, professor of obstetrics-gynecology and psychology at Wayne State University, noted that, "the belief that life had its origins in a single basic substance is so widespread among the various peoples of the world, primitive or civilized, that it can be considered one of the few universal themes in the history of ideas."[33] I believe that evolutionism was widely taught throughout the region of Babel. Henry Morris was convinced of this as well. Although lengthy, let me quote Morris on the religion of Babel, which was rooted in pantheistic evolutionism:

> The rebellion at Babel consisted not only of the people's refusal to scatter around the world, as God had instructed, but also of their instituting the new world 'religion' in the temple at the top of the Tower. The Tower had not been designed to 'reach unto heaven' in the physical sense (this would have been an absurd thing to attempt, as Nimrod and his colleagues well knew), but to reach heaven spiritually, there worshiping and communing with the 'host of heaven,' and to 'make us a name' rather than honoring the name of the true Creator.
>
> This host of heaven consisted of the sun-god, the moon-god, and the 'gods' represented by the various planets (Saturn, Mars, Venus, etc.), as well as the other stars. They actually involved the great hosts of rebel spirits that have fought against God and his saints all through the ages…
>
> It was almost certainly here that the Sumerian priests [the Sumerians were the original people of Babylon] were instructed in the secrets of astrology and the other occult sciences, as well as the religion of evolutionary pantheism…[34]

Evolutionary pantheism is the mystical form of evolutionism that invokes the impersonal Great Spirit in the sky, an imaginary god-force with no real power, of course. Pantheistic evolutionism is synonymous with its atheistic counterpart in the sense that both versions deny the one true – and very personal – Creator-God of the Bible. Pantheists want to cling to the spiritual, but without the one true Spirit revealed in Scripture. Pantheism is an idea that is nearly as old as humanity itself (Romans 1:25).

If a person is interested in creating a man-focused religion, pantheistic evolutionism is the way to go. As a worldview, atheism fails in that people are hard-wired to know that the spiritual realm exists. This is a belief that all of us inherently understand (Genesis 1:27; Ecclesiastes 3:11; Romans 1:20, 2:14-15).

---

[33] Henry Morris, "Pantheistic Evolution." https://icr.org/article/pantheistic-evolution/
[34] Ibid.

Since we are aware of the spiritual realm, we need to invoke an explanation for origins that includes some version of a divine power. However, rebellious and unregenerated man cannot accept the one true God as the Creator. Therefore, an alternative version of God – either the impersonal god-force of pantheism, or even the aloof and uninterested god of deism – must fit the bill. Since the god of deism is generally considered to be a personal god and not an impersonal god-force, it is preferable for many people to accept the impersonal 'Force' of pantheism. I believe that Morris was correct when he wrote that Nimrod's religion – which was almost certainly forced upon the rebellious population of Babel – adhered to pantheistic evolutionism in some manner. That idea makes sense.

As we will see, the evidence for evolutionism – whether atheistic, theistic, or pantheistic – is weak, which should make us wonder why it has been such a widely accepted idea throughout mankind's existence. I believe part of the reason lies in influential – and ungodly – people throughout history using it to convince others that they are nothing more than a cosmic accident. People who are confident that they are made in God's image, and are therefore equal before their Creator, do not easily enslave. That was one of the reasons why the early Communist leaders in Eastern Europe tried to abolish the belief in God. It is easier to rule over a people who lack that sense of God-given equality and rights.

Evolutionism and the Earliest Known Civilizations

Most civilizations in the ancient world that we know about today subscribed to an eternal universe, the spontaneous generation of life from non-life, and some type of 'upward' evolutionary scheme from simple to more complex life forms.[35] "For instance, from ancient observation it seemed that fleas originated from dust and maggots from cadavers. Aristotle was one of the first to teach this theory."[36] The Babylonian historian and astronomer Berosus, who was also a priest of Baal-Marduk, held to the belief "that life arose through a process of generation or evolution from simpler organisms."[37] Other, even earlier, examples include the Egyptians and the Hindu Brahmins. The Egyptians believed that "mice arose from the mud of the Nile,"[38] and the Hindu Brahmins taught that the universe was not only eternal but marked by endless cycles of expansion and contraction – a 'Big Bang' followed by a 'Big Crunch.'

Based upon a clear reading of Genesis, and from what we know about science today, why did these ancient civilizations hold to these beliefs? After all, both

---

[35] This is not to say that these ancient civilizations were purely atheistic in their worldview – they most certainly were not – but evolutionary ideas did make it much easier for later skeptics to propagate the idea that chance was responsible for the universe and everything in it.

[36] Benno Zuiddam, "Was Evolution Invented by Greek Philosophers?"
https://creation.com/https://www.creation.com/images/pdfs/tj/j32_1/j32_1_68-75.pdf

[37] Sibley, *Cracking the Darwin Code*, 9.

[38] David Menton, "The Origin of Life." https://answersingenesis.org/origin-of-life/origin-of-life-chance-events/

Genesis 1:1 and modern cosmology reveal that the universe had a beginning. Likewise, both Genesis and science reveal that life comes only from life, rather than having spontaneously generated from non-living matter. We also know from Scripture and science that everything has its own distinct genetic code ('after its kind'), which had to be created as information always comes from one source: A thinking mind. Why was it that not everyone in the ancient world believed in special creationism? Is it possible that there was, even millennia ago, a deception in the making?

In the ancient world, the notable exception to evolutionary thinking were the Hebrews, who believed in one supreme God who intentionally crafted the universe and everything in it (Genesis 1:1; Exodus 20:11; Psalm 19:1). How did the Hebrews know the truth about origins when all the nations around them taught evolutionism in some form? Clearly it was through divinely revealed knowledge.

Even though some of the pagan nations surrounding the Hebrews had attained a high level of civilization, their science was dismal at best. The Hebrews, through a reliance upon God's revelation, understood scientific truths more fully than those around them. Evolutionism, which permeated almost all the ancient world, offered no advantage to mankind whatsoever. Instead, evolutionism dramatically hindered the progress of science, technology, and medicine for not only centuries but millennia. This should lead us to a theological question: Would God want mankind to be intellectually capable or incapable? If capable, then we can help each other more effectively, and therefore we can truly become our 'brother's keeper' (Genesis 4:9). But if we are intellectually incapable and therefore unable to help one another, who benefits from that? God? Or Satan? Evolutionism is truly a 'hollow and deceptive philosophy' (Colossians 2:8) that is used by Satan and his followers to further crush fallen man.

Among other things, Genesis served as a corrective for the erroneous views of the pagan nations that surrounded the Hebrews. Moses, writing under the inspiration of God's Spirit, revealed several important truths about creation. First, the universe had a beginning. This point cannot be stressed enough: The Hebrew's were the only people in the ancient Near East, and possibly in the entire world, that believed the earth and universe had a beginning point. The other surrounding nations believed in a universe that had always existed.[39] Although people from ancient times to the present have believed in a universe without a beginning, the Bible always had it right: "In the beginning God created the heavens and the earth."[40]

Second, God – not the gods – created the heavens and the earth. All the nations

---

[39] The nations that held to the belief in an initial creation event of some kind still believed in a pre-existent material that was present before the beginning, most often being the primordial waters or 'the deep.' This is worlds apart from Genesis creation, in which nothing existed before the one true God spoke everything into existence (Genesis 1:3).

[40] Genesis 1:1, NIV.

surrounding the Hebrew's were polytheistic, believing in a god for this and a god for that. On the other hand, many of the great civilizations in the East viewed God as an impersonal creative force. Only the Hebrew's believed in the one true God revealed in the scriptures. All the other ancient Near Eastern cultures believed that the gods were simply an inherent part of the universe, shaping the creation out of eternally pre-existent matter. The ancient Hebrew's alone believed in creation *ex nihilo*.

Third, God created the universe in an orderly manner. God created in steps, the six days of creation. Undoubtedly, the six days of creation and the seventh day of rest form the basis for our seven-day week: Six days of work followed by the Sabbath, or 'holy day of rest.'

Fourth, molecules-to-man evolutionism is an impossible explanation of how the various life forms came into existence. Genesis 1-2 reveals that all the major categories of plant, animal, and human life were distinctly created by God. Vegetation (Genesis 1:11-12), the creatures of the sea and the air (Genesis 1:21), and the creatures of the land (Genesis 1:24-25) all came into existence, in their basic forms, directly from the hand of God. Likewise, humans were specially created by God, and are not evolved from pre-existing life forms (Genesis 1:26-27). One type of plant or animal does not eventually turn into another type of plant or animal. All the various living forms bring forth 'according to their kinds.'

Fifth, human beings are the pinnacle of God's creation. Humans were created in the 'image of God' (Genesis 1:27) and therefore are of great worth. This explains why we are to respect all human life. In contrast, the gods of the surrounding nations treated people in a capricious manner. These gods even used humans as slaves to carry out the menial work of tending to the creation. (Were these gods completely imaginary, or were there real entities behind each of them? Because of the spiritual warfare aspect of human history, I contend it is the latter.)

Moses offered a theological corrective for the surrounding nations. Yet evolutionism in its various forms persisted from the ancient world to the present. Even the prophet Jeremiah, whose ministry to the Hebrews began 600 years before Christ, may have discussed evolutionism in his time. Of the Hebrews who lived before him, at least some claimed that they were descended from the elements of nature: "They say to wood, 'You are my father,' and to stone, 'You gave me birth.'"[41] Was Jeremiah merely relying upon allegory in his writing, to make the point that many Hebrews denied the one true Creator? Possibly, but in the light of the evolutionism that pervaded the ancient world it seems there is more to this than just allegory.[42] Even God's people, throughout their history,

---

[41] Jeremiah 2:27, NIV.

[42] Although these idols made of wood and stone represented pagan gods, it is almost certain that these apostate Hebrews held to the idea that these gods fashioned people from the elements of nature, as described in the creation myths of Egypt, Mesopotamia, and other surrounding nations.

fell into the trap of pagan thinking at times.

Evolutionism in Ancient Greece

Early evolutionism, at least in a form that more closely resembles modern Darwinism, began to take shape near the beginning of Greek philosophy. The American paleontologist Henry Fairfield Osborn (1857-1935) noted his surprise when he discovered this truth: "When I began the search for anticipations of the evolutionary theory…I was led back to the Greek natural philosophers and I was astonished to find how many of the pronounced and basic features of the Darwinian theory were anticipated even as far back as the seventh century BC."[43] Just how original was Charles Darwin, anyway? When it comes to evolutionism, there truly are no new ideas (Ecclesiastes 1:9).

In ancient Greece, there were two schools of thought that constituted philosophy: Pantheism and materialism. The pantheists[44] maintained that God and the universe are essentially one and the same, while the materialists[45] taught that only matter (atoms and molecules) existed, with no God or gods overseeing the universe. In short, the pantheists believed in God as a natural force, while the materialists were atheistic in their worldview.[46] Both groups harbored wrong ideas about God and creation.

It is widely believed that the pantheistic philosophers began with Pythagoras, five and a half centuries before Christ. He traveled widely throughout the ancient world, "studying the esoteric teachings of the Egyptians, Assyrians, and even the Brahmins"[47] as well as the teachings of the Celts. The evolutionary ideas of India – an eternal universe marked by cycles of progression and regression as well as the evolution of lower-to-higher life forms – came to Greece primarily through Pythagoras. Like today, ancient ideas about God and creation did not operate in a cultural vacuum but were shared among the various people of the world.

The materialistic school of philosophy is generally considered to have begun with Thales of Miletus, six centuries before Christ. However, Church historian Benno Zuiddam maintains that none of the philosophers associated with the Milesian school were purely materialistic (atheistic) in their thinking. However,

---

[43] Henry Fairfield Osborn, *From the Greeks to Darwin* (New York, NY: Charles Scribner's Sons, 1929), xi.

[44] The Stoics, who were founded in the third century BC by Zeno of Citium, are the best example of those who held to pantheistic beliefs. Paul confronted Stoic philosophers while in Athens (Acts 17:16-34).

[45] The Epicureans, who were founded around 307 BC by Epicurus of Samos, are the best example of those who held to materialism. Paul also confronted Epicurean philosophers while in Athens (Acts 17:16-34).

[46] It must be noted, however, that pantheism and atheism are synonymous insofar that neither philosophy holds to the belief in a personal God – and God without the attributes of personality is meaningless.

[47] Paul James-Griffiths, "Exposing the Roots of Evolution." Unmasking Fables, Promoting Truth DVD (Creation Ministries International). https://usstore.creation.com/product/401-unmasking-fables-promoting-truth-dvd James-Griffiths is quoting Iamblichus, Pythagorus' biographer.

"it can be argued that their philosophies contain building blocks that, as such, are also used in modern evolutionary concepts."[48] The basic ideas of evolutionism were put into place by these ancient thinker's millennia ahead of our time, even though they may not have been atheists themselves. As far as anyone can tell, the members of the Milesian school held to a belief in the spiritual realm, which seems to have made them the originators of a primitive form of theistic evolutionism. It seems that the so-called 'atomists' – primarily Democritus, but others as well – were the first thinkers to postulate a clear-cut materialism, with no "Great Principle"[49] or God overseeing creation. They seem to be the first undisputed proponents of the materialistic worldview.[50] We will look at Democritus later, but first let us continue investigating Thales and his school in Miletus.

Thales is often considered to be the "father of Western philosophy"[51] who "founded the Milesian school of natural philosophy."[52] Like Zuiddam, creationist Bill Cooper notes that "it is very doubtful that Thales was a materialist [atheist] at all"[53] as based upon some strong creationist-type statements that are attributed to him. Regardless of whether Thales was devoted to pure materialism or not, the evolutionary beliefs of his student Anaximander would lead one to believe that Thales taught some form of evolutionism.[54]

---

[48] Zuiddam, "Was Evolution Invented by Greek Philosophers?" Although the philosophers that Zuiddam addresses in this paper – Thales, Anaximander, and Empedocles – are shown not to be evolutionists in the specifically Darwinian sense of the term, they nonetheless held to a primitive form of evolutionism.

[49] Allan Chapman, Slaying the Dragons: Destroying Myths in the History of Science and Faith (Oxford, England: Lion Books, 2013), 27.

[50] It should be noted, however, that the Epicureans did allow for the existence of the gods, but they were composed purely of matter and were aloof and unconcerned with the affairs of men. This was a vastly different view of the gods than the general population of Greece held.

[51] R.C. Sproul, The Consequences of Ideas (Wheaton, Illinois: Crossway Books, 2000), 15.

[52] David Menton, "The Origin of Evolutionism: It Didn't Begin with Darwin."
https://answersingenesis.org/theory-of-evolution/origin-of-evolutionism-didnt-begin-with-darwin/

[53] Bill Cooper, After the Flood (West Sussex, England: New Wine Press, 1995), 21.

[54] That is, unless Anaximander came up with the tenets of evolutionary thinking all on his own. However, the experience of history demonstrates that students tend to follow very strongly in the footsteps of their mentors. If Thales believed in a Supreme Being of some kind as well as the idea of evolutionism, this would have made him an ancient proponent of theistic evolutionism. Thales may have held to a form of polytheistic evolutionism, in which the gods fashioned the pre-existent matter of the universe into everything we see in nature. Based on the rampant polytheism in the Greek culture, this seems to be the likely case.

ANAXIMANDER

Anaximander seems to be the first 'anti-creationist,' as it is noted that, "it is to him that we must look for the first recorded challenge to creationism from the materialistic school."[55] He taught that the first animals to appear were marine creatures, which is correct according to both Genesis and secular scientists, but he also taught that these creatures formed spontaneously out of mud – clearly an incorrect assertion. Although the idea of the spontaneous generation of life from non-life was shown to be wrong in the sixteenth century by the Italian physician Francesco Redi and confirmed to be in error by the French biochemist Louis Pasteur three centuries later, evolutionists today must still hold to some form of spontaneous generation or "abiogenesis."[56] Anaximander maintained that these marine animals in turn gave rise to land animals, birds, and eventually human beings. This is without doubt evolutionism at its finest – or worst, depending upon one's viewpoint!

Two centuries later Empedocles proposed similar evolutionary ideas. The prolific creationist author Jerry Bergman notes that, "The Greek philosopher Empedocles (493-435 BC), often called the father of evolutionary naturalism, argued that chance alone 'was responsible for the entire process' of the evolution of simple matter into modern humankind."[57] That is no different than the ideas of modern evolutionists, of course. Bergman continues: "Empedocles concluded that spontaneous generation fully explained the origin of life, and he also taught that all living organisms gradually evolved by the process of trial-and-error recombinations of animal parts."[58] All naturalistic evolutionists today must subscribe to spontaneous generation having happened at least once in the very distant past, as well as some form of 'upward' evolution from lower-to-higher

---

[55] Cooper, 21. As with Thales, it seems that Anaximander also subscribed to a universe imbued with supernatural forces. However, the god of Anaximander was certainly not the one true God of the Bible, but like many Greeks Anaximander likely confused the Creator with the creation, something that Paul warns us against (Romans 1:25).

[56] Henry Morris, The Long War Against God, 207.

[57] Jerry Bergman, "Evolutionary Naturalism: An Ancient Idea."
        https://creation.com/evolutionary-naturalism-an-ancient-idea

[58] Ibid.

life forms. As we have seen, those assumptions were no different in the ancient world.

Even though Empedocles lived over two millennia prior to the time of Darwin, many of Darwin's ideas were already well-established through Empedocles' teachings. A century after Empedocles, Aristotle (384-322 BC) continued to teach that all life forms evolved from simple-to-complex by means of a self-organizing biological process. Although this idea has been proven impossible by modern laws of science, it nonetheless held sway throughout the ancient world.[59] The ideas of Anaximander, Empedocles, and Aristotle are not all that different from the evolutionism of today.[60]

The Milesian school of natural philosophy may not have been thoroughly materialistic in its views, as some suppose, but that all changes dramatically with Democritus and the 'atomist' school of philosophy. Democritus (460-370 BC) is the so-called 'father of the atomists.'[61] He claimed that there was no purpose to the world, and there was no Supreme Being or 'First Cause' who was responsible for its existence. For Democritus, 'Ultimate Reality' was not God or even the gods, but matter (atoms and molecules). He believed that everything is purely material in nature, with no spiritual realm to be concerned about. Democritus set the stage for the next major thinker in materialistic evolutionism, Epicurus. Even today, Epicurus is associated with the idea of godless (undirected) evolutionism.

Continuing in the long line of Greek evolutionists, Epicurus (341-270 BC) "taught that there was no need of a God or gods, for the Universe came about by a chance movement of atoms."[62] Atheism is clearly not a modern idea, as some suppose! It is no wonder that Paul, when confronting the ideas of the Epicurean philosophers in Athens, began his speech with the doctrine of creation *ex nihilo* (Acts 17:24-27). If one does not believe in the truth of Genesis 1:1, then there is no reason for him or her to believe in the truth of John 1:1 either.

Evolutionism in Ancient Rome

By the time the Roman Empire was the ruling power in the known world,[63] the evolutionary beliefs of Epicureanism dominated Greco-Roman thinking.[64] In

---

[59] Namely the principle of entropy, as described by the Second Law of Thermodynamics.

[60] At least in the sense that evolutionary changes are attributed to chance, random processes in nature, as opposed to God creating with intentionality. We would not expect the ideas of Empedocles to be exactly the same as the concepts described in Darwin's *Origin of Species*, at least in specific terms, but in a general sense we do see the broad concept of purely naturalistic change over time in the work of both men.

[61] Although Democritus is often referred to as the founder of the atomists, it must be noted that he closely followed the teachings of Leucippus.

[62] Paul James-Griffiths, "Evolution: An Ancient Pagan Idea." https://creation.com/evolution-ancient-pagan-idea

[63] Rome became an official empire in 27 BC, when Octavian became the first emperor. However, it had become increasingly powerful well before that time.

[64] In the time of the apostles, the three schools of philosophical thought that dominated the Gentile culture were the Epicureans, Stoics, and Platonists. Interestingly, Paul encountered the

fact, some theologians today still refer to naturalistic evolutionism as 'Epicureanism.'[65] Lucretius (99-55 BC) was a Roman philosopher who kept Epicurus' teachings at the forefront of Greco-Roman thought. He taught Darwin's "essential ingredients"[66] two millennia before Darwin, promoting the idea that "the earth had spontaneously generated all living forms."[67]

The poet Ovid (43 BC-AD 17) continued to promote the idea that the earth spontaneously generated all living forms. Ovid accepted the spontaneous generation of the Egyptians, Greeks, and Lucretius before him:

> For when moisture and heat unite, life is conceived, and from these two sources all living things spring. And, though fire and water are naturally at enmity, still heat and moisture produce all things, and this inharmonious harmony is fitted to the growth of life. When, therefore, the earth, covered with mud from the recent flood, became heated up by the hot ethereal rays of the sun, she brought forth innumerable forms of life; in part she restored the ancient shapes, and in part she created creatures new and strange.[68]

Fortunately, for every Epicurus, Lucretius, or Ovid who opposed divine design in nature, there was a Cicero (106-43 BC) who fought back. Although Cicero was a Stoic philosopher and not a biblical creationist,[69] he nonetheless challenged "the evolutionary ideas of the philosophers of his day"[70] for "as a creationist, the Stoic Cicero simply could not appreciate the Epicurean viewpoint of Lucretius,"[71] his contemporary and philosophical opponent.

Epicureans and Stoics directly while in Athens (Acts 17:16-34), and he was undoubtedly aware of the teachings of Plato and his followers as well.

[65] For example, N.T. Wright, in *Surprised by Scripture*, associates modern, naturalistic science with Epicureanism.

[66] Donald D. Crowe, *Creation Without Compromise* (Brisbane, Australia: Creation Ministries International, 2009), 23-26.

[67] Louis Lavallee, "The Early Church Defended Creation Science." https://icr.org/article/early-church-defended-creation-science

[68] Ovid, *Metamorphoses* (Book 1, "Other Species are Generated"). https://ovid.lib.virginia.edu/trans/Metamorph.htm#488381088

[69] Cicero lived decades before the time of Christ and the writings of the apostles and early Church leaders. Although he would have been aware of the Jewish scriptures, there is no reason to believe that he based his worldview upon them.

[70] Russell Grigg, "A Brief History of Design." https://creation.com/a-brief-history-of-design

[71] Cooper, 29.

CICERO

Cicero – undoubtedly the greatest statesman, lawyer, and philosopher of his day – voiced a strong case for divine design eighteen centuries before William Paley penned *Natural Theology* in 1802, which is still considered by some to be the 'gold standard' for describing divine design in nature. Cicero "reasoned that the movement of a ship was guided by skilled intelligence, and a sundial or water clock told the time by design rather than by chance. He said that even the barbarians of Britain or Scythia could not fail to see that a model which showed the movements of the sun, stars, and planets was the product of conscious intelligence."[72] Due to his stance on divine creationism and his insistence upon maintaining a strong code of ethics, Cicero was referred to by some in the early Church as a 'righteous pagan.' (This may seem to us today as being a sort of back-handed compliment!) Cicero's design argument has never been refuted, nor has it gone out of fashion, simply because it appeals to a common-sense interpretation of nature.

Decades later the apostles would articulate the same position on intelligent design as Cicero did, but this time with the assistance of God's Spirit (2 Peter 1:20). They did this for the benefit of posterity, of course, but also because even in their day there were skeptics such as Pliny the Elder (AD 23-79), who wrote, "We are so subject to chance that Chance herself takes the place of God; she proves that God is uncertain."[73] The element of chance in creation had already been present in Greco-Roman thinking for a long time, and it is still a deceptive idea today – truly it is a formidable 'doctrine of demons.'

Adam, Human Evolutionism, and the Antiquity of Man

Creationists fight over the age of the earth. That is not exactly a news flash for most Christians. In my twenty-five plus years as a believer, I have been both a young earth and an old earth creationist. Eventually, after much frustration with the issue of the earth's age, I declared myself a 'general creationist' during the time that I wrote my first book. When preparing *Worldviews in Collision*, I utilized resources from both camps, trying to discern to the best of my ability

---

[72] Grigg, "A Brief History of Design."
[73] James-Griffiths, "Evolution: An Ancient Pagan Idea."

which view makes the most sense in different areas. That made me a man without a country – and that is often a hard situation to be in. We all want to feel surrounded by like-minded people who embrace us as one of their own.

Today I consider myself a creationist, period. I am a recent creationist, because I feel that this position makes sense of Scripture when read in a straightforward manner. But I will not attack others for their opposing beliefs. If Ken Ham came to my city for a creation talk, I would be the first person in the door to welcome him, and I can say the same thing about Hugh Ross. I would eagerly listen to both men, openly considering what they have to say about origins. Unlike some apologists today, I have no problem with views that oppose mine. I do not know everything there is to know about science and theology, so I am learning to keep my mouth shut and my ears open. That, by the way, is not an easy trick for fallen man – and especially for me.

Having said that, however, I strongly suspect the creation is older than 6,000 years. I do not believe the creation has to be millions or billions of years old, or even hundreds of thousands of years old for that matter. Just a little older than 6,000 years. Most people who question, doubt, or even deny a young creation do so because of cosmology – the 'distant starlight problem' is almost always the culprit – but I have no problem with a young universe that is billions of light years across. We really know little about cosmology. For all we know, the speed of light and the expansion of space during creation week may have been much greater than it currently is. If we add to that a host of other factors, I can see how the universe can be both young and billions of light years in diameter at the same time. Young earth creationist astronomers have proposed several intriguing models for a recent creation that is 93 billion light years in diameter.[74] I also fail to be convinced by radiometric dating as conclusive evidence for an old earth. Like cosmology, it relies upon several starting assumptions that should be seriously questioned. Nonetheless, I do question only 6,000 years of history. I feel like we need just a few more thousand years for everything to make sense. That still makes me a recent creationist, but there will be some who will question my sincerity or even my faith because of that. I briefly wrote about the age debate in the introduction, but I will discuss it in much more detail in chapter six.

As the years have gone by, I find myself less likely to 'go for the jugular' regarding creationist views contrary to mine. Despite the claim of some recent creationists, old earther's are not always 'compromisers.' Two of the most amazing Bible teachers-theologians-apologists I know are old earth creationists – and I know no one who cherishes' and respects God's Word more than they do. On the other hand, young earther's are not 'unenlightened imbeciles' who should have lived in Elizabethan times, a sentiment shared by some old earth proponents. Some of the most amazing creationists I know hold to the young earth position. We must learn to emphasize the 'with gentleness and respect' part of 1 Peter 3:15 more than ever, if we are going to become winsome apologists.

---

[74] This is the current estimate for the size of the observable universe.

In the words of the late Rodney King, "Why can't we all just get along?"[75]

There is an old Christian tradition concerning the Apostle John. John was known as the 'Apostle of Love' for good reason: Love forms the basis of his letters to the Church (1-3 John). John was a great defender of truth, upholding the teachings of Jesus despite the onslaught of false teachings that took place in his day, when the Church had barely begun. John opposed false teachers such as Cerinthus only because he had great love for the truth – and truth can only be found in Christ and his teachings.

The tradition tells of the aged John teaching about the most important thing in the life of a Christian believer: Learning to love others. As the story goes, near the end of John's life he was frequently asked to share about what matters most in the life of believers, and his words were always the same: "Little children, I tell you to love one another." Eventually a pastor or some other church leader said to him, "Brother, I think we have already heard that. You walked side-by-side with Christ and learned from him firsthand. Do you have any other teachings from the Lord?" John answered this fellow with some of the greatest words never recorded in Scripture: "If, one day, you can understand this teaching, then there is nothing else you need to learn."

It is difficult to know the source behind that story for sure. Maybe it originally came from Polycarp or another disciple of John's, and then was later relayed to Irenaeus or another leader in the Church. Or perhaps one of the great Church historians heard the story and recorded it for the sake of posterity. Regardless of the source, however, the story applies to us today just as strongly as it did to those who may have heard it directly from John himself. Sometimes we need to set aside complex theology and focus on love and respect for others. For John, in the end all that really mattered was that Christian believers learned to love one another. Sometimes creationists from both camps need to put this into practice as well. I have certainly been guilty of this a time or two.

There is one area related to the 'age issue' that I am especially adamant about, however: The doctrine of a real, historical Adam and his recent creation. Moses, writing under the inspiration of God's Spirit, tells us in no uncertain terms that Adam was specially created by the Lord near the end of creation, on the sixth day (Genesis 1:26-27). Jesus spoke of a real, historical Adam (Mark 10:6), and chapter fifteen of 1 Corinthians – the resurrection chapter of the New Testament – makes no sense without a real Adam in time and space. Yet so many theologians today are insisting that Adam was merely a symbolic figure, representing the whole of humanity. For example, the theistic evolutionist Denis Lamoureux believes that Adam never existed. That idea flies in the face of what Moses, Jesus, and Paul believed. Dr. Lamoureux also contends that Adam's non-existence has no impact on the foundation of the Gospel message. Regarding the Gospels, he writes that "there is no mention whatsoever of Adam and whether or not he existed. Christian faith is founded on Jesus, not Adam…We must also

---

[75] Although it is debatable if King spoke those exact words.

separate, and not conflate, the historical reality of Jesus and His death and bodily resurrection from the fact that Adam never existed."[76] I have no doubt that Dr. Lamoureux is a devout believer. I say that because it seems to be common for some creationists to doubt the salvation of those who do not believe exactly as they do. I, however, will not engage in that sort of arrogant nonsense. I have no reason to doubt that Dr. Lamoureux sincerely confesses Christ as his Lord and Savior (Romans 10:9), despite not believing in a literal Adam. A person can be wrong about any number of theological issues and still be part of the Christian body of believers. Our salvation does not depend upon us passing a theological test. But we really do need a real, historical Adam to make sense of the Gospel message.

The English creationist Simon Turpin, who excels in the theological aspect of creationism, writes the following:

> The death and resurrection of Christ are the central events of the gospel. But notice that Paul says that Christ died for our sins. Why did Christ have to die for our sin? In his letter to the Romans, Paul argued that it was through one man that sin entered the world, and death through sin (Romans 5:12-16). Christ had to die on the Cross because Adam had brought sin and death into the world. Therefore, just as God provided an atonement for Adam's sin (Genesis 3:21), so Christ has provided atonement for our sins because we are "in Adam" (1 Corinthians 15:22). Christ's death was an atoning sacrifice for sin.[77]

The Gospels make no sense to me without a literal Adam. Jesus could not be the 'Second Adam' if the 'First Adam' was just a myth. If Adam were merely symbolic as the 'First Adam,' this would imply that Jesus was also symbolic, as the 'Second Adam.' We really do need to insist upon the historicity of Adam.

A related issue with a significant theological impact involves dating Adam's creation. Creationists from both young and old earth camps insist upon a real Adam only recently created, whereas evolutionists are convinced that vast ages have passed in the development of modern man from a primitive (non-human) ancestor. The idea of extremely ancient origins for mankind goes back at least to the time of the Babylonians, Egyptians, and Hindu Brahmins. Recent creationists date Adam at no more than 6,000-10,000 years ago, while old earth creationists generally put the figure at "10 to 30 thousand years ago, but perhaps as late as 60,000 years ago."[78] Although I personally side with the former camp in this

---

[76] Simon Turpin, "How Do Some Among You Say There Is No Adam?"
  https://answersingenesis.org/adam-and-eve/how-do-some-among-you-say-there-no-adam/

[77] Ibid.

[78] Jon Greene, "A Biblical Case for Old-Earth Creationism."
  http://godandscience.org/youngearth/old_earth_creationism.html#when

issue, both creationist groups believe that Adam was created long after the dates claimed by the ancient civilizations just mentioned. The creationist Paul James-Griffiths notes that, "Some of the Babylonians claimed that they had astronomical inscriptions on clay tablets for 730,000 years. Others, like Berosus, claimed 490,000 years for the inscriptions. The Egyptians claimed that they had understood astronomy for more than 100,000 years."[79] This idea of a very ancient origin for mankind was common with the Greeks as well: "Plato and many others of the philosophers, since they were ignorant of the origin of all things, and of that primal period at which the world was made, said that many thousands of ages had passed since this beautiful arrangement of the world was completed."[80] Perhaps these great dates among some of the ancient nations is why Paul, when addressing the issue of false teachers with Timothy, notes that the heretical teachers within the Church often resorted to passing on "endless genealogies"[81] to the naïve and unsuspecting.

It is wise to consider these words from the early Christian apologist Theophilus, writing to his non-believing friend Autolycus only a century and a half after the time of Christ:

> For my purpose is not to furnish mere matter of much talk, but to throw light upon the number of years from the foundation of the world, and to condemn the empty labor and trifling of these authors, because there have neither been twenty thousand times ten thousand years from the Flood to the present time, as Plato said, affirming that there had been so many years; nor yet 15 times 10,375 years, as we have already mentioned Apollonius the Egyptian gave out; nor is the world uncreated, nor is there a spontaneous production of all things, as Pythagoras and the rest dreamed; but, being indeed created, it is also governed by the providence of God, who made all things; and the whole course of time and the years are made plain to those who wish to obey the truth.[82]

What is 'made plain' can be found within the genealogical records of Genesis, showing a relatively recent creation date for Adam. Even if gaps exist in the Old Testament genealogical records, we can say with certainty that human history is brief compared to evolutionary claims for mankind's antiquity. The biblical history of mankind is nowhere near long enough to support the idea of human evolution, which requires not just hundreds of thousands but millions of years.

---

[79] James-Griffiths, "Evolution: An Ancient Pagan Idea."
[80] Ibid.
[81] 1 Timothy 1:4, NIV. Admittedly, there is speculation behind this idea for the meaning of 'endless genealogies,' but it does make sense in the light of earlier pagan teachings concerning the great antiquity of mankind.
[82] Theophilus, *To Autolycus* (Book 3, Chapter 26). https://newadvent.org/fathers/02043.htm

If theistic evolutionism were true, there were countless 'pre-Adamites' on the earth, extending back hundreds of thousands of years. Theologically, this is a problem. If there were beings who were anatomically similar if not identical to Adam, but not made in God's image as Adam was, what would happen to them when they died? For that matter, how would their DNA – if it still exists in the human genome – spiritually affect us today? Scripture makes it clear that Adam was the first man (Genesis 1:26-27; Mark 10:6). There is no scriptural evidence for people existing prior to Adam and Eve, and the supposed scientific evidence is highly suspect. The ancient idea that there were people on the earth for hundreds of thousands of years aligns with evolutionism, but not creationism. As we have seen, the debate over man's origins is very ancient, going far back into history. Nonetheless, it is still exceedingly relevant today.

'There is Nothing New Under the Sun'

Even by the conclusion of this first chapter, we can clearly see that evolutionary ideas have a long history – far longer than many of us were previously aware of. "All of the seeds for acceptance of long ages and belief in an evolutionary progression were present in ancient times; being derived from pagan beliefs as far afield as India, Egypt, Babylon, and Greece."[83] Evolutionary thinking is a worldview, a religious philosophy, even a 'doctrine of demons' intended to shift the focus away from the one true God of the universe. It is truly spiritual warfare in every sense of the term.

Sadly, these evolutionary ideas are alive and doing very well today. Despite Scripture showing itself to have originated from God (2 Timothy 3:16; 2 Peter 1:20), the battle between evolutionism and creationism is still raging on after all these centuries. As Solomon wrote three thousand years ago, "there is nothing new under the sun."[84]

In the next chapter, we will see that the early Church understood the importance of the origins debate as an apologetic for an unbelieving world. Some of the greatest apologists to ever live were part of the early Church. Names such as Justin Martyr, Clement of Alexandria, Origen, and Augustine defended Genesis creationism in the face of pagan evolutionism. But no one defended biblical creationism better than Paul. His address to the philosophical elites in Athens (Acts 17:16-34) is still the template for sharing creation and the Gospel with unbelievers. It is to these earliest Christian apologists that we turn to next in our study.

---

[83] Sibley, *Cracking the Darwin Code*, 16.
[84] Ecclesiastes 1:9, NIV.

CHAPTER TWO

The Early Church Confronted Evolutionism

Christian apologetics began when the Church was born. There has never been a time in Church history when apologists were not fully engaged in defending and proclaiming the faith. Among the topics that apologists specialized in, the origins debate has always been near the top of the list. The apostles and the Church Fathers knew the importance of the creation message: If one gets the issue of origins wrong, everything else of importance – especially salvation – will be answered incorrectly as well. To better understand the evidence for biblical Christianity today, it is beneficial for us to know what the early Church taught about God, creation, and the big questions of faith.

The Church formally began after the ascension of Christ sometime between AD 30-33, on the day of Pentecost as described in Acts 2:1-12. The Church Age began a whole new dimension of debating the origins issue, as some of the greatest champions of biblical creationism wrestled against the skeptics of their time. The apostles Peter (2 Peter 3:3-7), John (John 1:1-3; Revelation 4:11), Paul (Acts 17:24-26; Romans 1:20; Colossians 1:16-17), and the unknown author of Hebrews (Hebrews 1:1-2) all wrote of God's special creation, while their opponents included Cerinthus and many others of a similar Gnostic disposition who sought to dispute them at every turn.[85]

John used the Greek philosophical term for 'principle of universal reason' that was used to denote the 'Ultimate Reality' behind the universe.[86] John wrote, "In the beginning was the Word (*logos*), and the Word (*logos*) was with God, and the Word (*logos*) was God,"[87] and, "That which was from the beginning, which we have heard, which we have seen with our eyes, which we have looked at and our hands have touched – this we proclaim concerning the Word (*logos*) of life."[88] John likely chose this word, *logos*, as a way to connect with the Greek-speaking, philosophically minded Gentile crowd. John was attempting to convince his non-Jewish readers that belief in Jesus as God incarnate is reasonable. Like his friend and fellow apostle Paul, John sought to forge a bridge to the non-Jewish world, which was steeped in evolutionary paganism.

In similar fashion, Clement of Rome – who co-labored in ministry with Paul (Philippians 4:3) – "deliberately chose to describe the God of the Bible as the

---

[85] Although there were naturalists who opposed biblical creationism in the time of the apostles and the early Church, it was more common for the skeptical to possess a Gnostic worldview. Simon Magus, Menander, Basilides of Alexandria, Marcion of Sinope, Apelles, and Valentinus serve as examples of Gnostics opposed to a straightforward reading of Genesis creation.

[86] Heraclitus is often credited with being the first to use the term *logos* as a connection to God or 'Ultimate Reality.'

[87] John 1:1, NIV.

[88] 1 John 1:1, NIV.

great Demiurge (Greek, *megas demiourgos*),"[89] in 1 Clement, a letter he wrote to the struggling Church in Corinth. This was the same term that Plato used to describe the 'Master Designer.' Through his carefully chosen words, Clement also sought to build a bridge to the philosophically minded Gentile world, demonstrating that Jesus is the personal Creator-God of the universe. Both Justin Martyr (second century) and Augustine (late fourth-early fifth centuries) would utilize the Greek term *logos* in their apologetic as well. Like us today, Clement, Justin, and Augustine fought against both the materialism and Gnosticism that threatens to diminish God's sure words concerning origins.

Paul: Confronting Evolutionism in First Century Athens

Perhaps the best example of an apostle in the first century battling against Greco-Roman evolutionism is found in Paul's speech to the Athenian philosophers (Acts 17:16-34), which was one of only two speeches before a pagan or non-Jewish audience that was recorded by Luke in Acts.[90] This speech is considered to be "the exemplary meeting between Jerusalem and Athens."[91] Paul's speech truly serves as the connection point in the battle between Judeo-Christian revelation ('Jerusalem') and Greek rationalism ('Athens'), a battle which continues to this very day in the form of evolutionism versus creationism.

PAUL IN ATHENS

Since it is imperative that we demonstrate the living power of the Gospel with others, "it is crucial to have a biblically based and carefully honed apologetic methodology in place before confronting the learned paganism of our age [evolutionism]."[92] There are still 'Athenian philosophers' alive and well today,

---

[89] David Herbert, *The Faces of Origins* (London, Ontario: D & I Publishing, 2004), 14.

[90] The other speech being found in Acts 14:15-17.

[91] Ron Vince, "At the Areopagus (Acts 17:22-31): Pauline Apologetics and Lucan Rhetoric." https://mcmaster.ca/mjtm/4-5.htm

[92] Kim Riddlebarger, "For the Sake of the Gospel: Paul's Apologetic Speeches." https://modernreformation.org/default.php?page=articledisplay&var1=ArtRead&var2 =445&var3=main

and they deserve a rational-yet-biblical response from Christian believers, for Christ himself commanded us to take the Gospel to the world (Matthew 28:16-20; Mark 16:14-18) and that includes those who are skeptical and maintain vastly different views regarding God and mankind's place in nature. Unfortunately, many Christians today are not prepared to engage the skeptics of our world – which is why apologetics is needed more than ever before.

Although Athens was steeped in pagan idolatry, Paul not only contained the displeasure that he surely felt, but he even went one step further and used that idolatry as a steppingstone into his discourse on the biblical worldview. As evangelists today, believers must also find points of contact with unbelievers that allow us to dialogue with them, capturing their interest long enough to make a proper presentation of the Gospel.[93]

Even though some have claimed that Paul was out of his league when confronting the intellectual elites in Athens, this was not the case at all. Before Paul was known as the Apostle Paul, he was first known as Saul of Tarsus, revealing to us the city of his birth (Acts 22:3). Tarsus was praised for being a major center of education, especially in philosophy and rhetoric, and was particularly known for its schools devoted to Stoic philosophy. Although Tarsus was one of three major cities known for its excellence in the study of philosophy and rhetoric – the other two being Athens and Alexandria – it was commonly held that Tarsus surpassed even these other great cities in this regard. As a Roman citizen almost certainly born into some wealth (Acts 22:22-29), Paul may have been privileged enough to have taken advantage of some of these educational opportunities and may have known far more about the Epicurean and Stoic philosophies than we realize. Perhaps an extensive education contributed in part to why Paul was chosen to be the 'Apostle to the Gentiles' (Romans 11:13). As Christians who have been commanded to take the Gospel to the world, we should also strive to educate ourselves concerning the beliefs of others.

The Epicurean philosophy was founded upon the atomic theory of Democritus. Atomic theory maintained that the universe consists of atoms or 'building blocks' which are eternal and constantly forming new combinations, giving rise to new physical forms over time. These new combinations of atoms are due to chance, in which atoms fall through infinite space and blindly interact with one another. This is clearly naturalistic evolutionism, with chance being the hero in this philosophy. According to the Epicureans, even the gods are ultimately constructed of atoms. Therefore, for the Epicurean atoms and molecules is what constituted 'Ultimate Reality.' The Epicureans were the

---

[93] 'Points of contact' should not be confused with 'common ground,' as the biblical worldview has little in common with the secular worldview in terms of areas where both agree. However, points of contact are common interests that both believers and unbelievers alike find intriguing, such as the issue of origins. Even though Genesis 1:1 is worlds apart from the naturalistic view of origins, the beginning of all things is nonetheless a topic that both groups find fascinating and worthwhile – and, therefore, a possible 'point of contact.'

naturalistic evolutionists of their day.

Although the Stoics believed in a creator, their creator was worlds apart from the Creator declared by Paul. The god of pantheism is nothing more than the impersonal embodiment of the laws of nature, and the Stoics, like all pantheists, confused the creation with the Creator (Romans 1:25). Paul, boldly declaring the one true God of the universe, proclaimed the personal God who is distinct from the cosmos, and who created everything from nothing.

Both groups of philosophers would have needed a crash course in creation-sin-redemption before they could contemplate the truth of Christ. Unlike Peter in his speech at Pentecost (Acts 2:14-41), Paul first had to establish the case for the existence of a personal Creator who is distinct from the universe, and intimately concerned for the affairs of human beings. In doing this, he effectively refuted the scientific materialism, polytheistic deism, and godless pantheism that collectively formed the philosophical background of Athens. Could Paul do this using Scripture, or would he need to address this crowd in a manner that relied more upon God's general revelation through creation, at least in the beginning of his speech? Paul did just that, first emphasizing the rationality of creation while eventually working the distinctly biblical doctrines of sin and redemption into his Gospel presentation.

Without the foundation of the one true Creator, non-Christians of all varieties can never really understand the message of the Gospel: "A gospel without the message of the Creator, and the origin of sin and death, is a gospel without the foundational knowledge that is necessary to understand the rest of the gospel."[94] One can never really grasp the message of the Gospel, such as the origin of sin and death, the need for redemption, and the restoration of all things, without first coming to know God as the transcendent Creator of everything. With that foundation in place, which is built up through the early chapters of Genesis, one can then begin to see the need for Christ. Paul realized this, and he began his Areopagus presentation 'in the beginning,' exactly where he needed to start.

Knowing that his audience is confused regarding origins, Paul began with these words: "The God who made the world and everything in it is the Lord of heaven and earth and does not live in temples built by human hands."[95] Creation is the ideal starting point when engaging those of non-Christian worldviews, as they are almost always steeped in some version of evolutionism. Before they could even begin to grasp the concept of Christ as the resurrected redeemer of fallen man, these philosophers first needed to understand Christ as the personal Creator of everything. Creationism must come first, so that rebellion and redemption can later be understood. Once evolutionism falls by the wayside, everything else begins to make sense.

Athens boasted many grand temples, most dedicated to a Greek god or

---

[94] Ken Ham, "Evangelism for the New Millennium."
        https://answersingenesis.org/gospel/evangelism/evangelism-for-the-new-millennium/
[95] Acts 17:24, NIV.

goddess, but Paul informed his listeners that God is not confined to buildings made by human hands. This idea was foreign to the Greek mind, however, as they were steeped in a polytheism that dedicated numerous temples to the gods as places for them to reside. Paul, as a Jewish Christian, recognized that even the one true temple in Jerusalem was constructed out of respect and worship of God, but did not contain or confine God in any way. Solomon, the builder of the first great Jewish temple, declared that the temple could not contain the Creator (1 Kings 8:27), and Stephen, much later in the early Christian era, reiterated this point as well (Acts 7:48).

In opposition to the Greek gods, Paul makes it clear that the one true God of the universe does not need to be served like a proud emperor: "And he is not served by human hands, as if he needed anything. Rather, he himself gives everyone life and breath and everything else."[96] God is independent and does not require anything from mankind for his existence. God is pleased when people offer true worship, but God does not need this worship for his existence, for God alone is self-existent. In comparison, the Greek gods thrived on the worship that men gave them.

By stating that God 'gives life and breath and everything else,' Paul was stressing God's role not only as Creator, but as Sustainer of everything. This was a concept that was foreign to the Epicureans, as they were atheists. The Stoics were much more comfortable with the idea of a god who sustains the creation, but their sense of creationism was pantheistic – everything in the universe merely emanated from their impersonal god-force. Their god was not even close to the God of Christianity, who is distinct from the creation and responsible for its existence.

Paul then stresses the sovereignty of God in human history: "From one man he made all the nations, that they should inhabit the whole earth; and he marked out their appointed times in history and the boundaries of their lands."[97] Although the Jews and Christians of the time would have thought of Adam when Paul referred to the 'one man' from which all others are descended, the idea of a single man as the progenitor of the human race was a foreign concept to the Greeks, who viewed people as being the physical offspring of a variety of gods who came to the earth in the distant past. Additionally, the notion that there is a sole, supreme Creator who watches over our lives, and even places us in the exact time and place that we should live, was a radical idea for the Greeks, whose worldview often emphasized chance happenings.

By proclaiming the unity of all human beings, Paul implied that all people are equal. All people are created in the image of God (Genesis 1:27). This was a major stumbling block for the Greeks: They thought of themselves as being superior to other civilizations, perhaps on account of their great accomplishments in philosophy and civics. Both groups of philosophers tended to attract the

---

[96] Acts 17:25, NIV.
[97] Acts 17:26, NIV.

learned elite of their culture, so this was likely shocking to hear from Paul. However, unity and equality are a major message of Paul's theology, which is summarized well in Galatians: "There is neither Jew nor Gentile, neither slave nor free, nor is there male and female, for you are all one in Christ Jesus."[98] Paul must firmly establish this point if he is to be successful at presenting the risen Christ later in the speech. It may be that many of those in attendance who scoffed at Paul did so because of this difficult teaching. The various worldviews represented in the Areopagus on this day would have come into serious conflict with Paul at this point.

Paul utilized a three-step approach to sharing his faith across worldviews. First, Paul initially sought to contrast the Hellenistic worldview with that of Scripture. Rather than seeking neutral ground, which does not exist, Paul tells his audience point blank that the two worldviews are in stark contrast to one another. Essentially, Paul is preparing them to make a choice: Accept the biblical worldview, or instead continue to walk along the path of falsehood. The situation is no different today, of course.

Second, Paul drives home the point that every person must either align with God's will or continue to be opposed to it. Paul was warning his audience that they can accept Christ now, as Judge and Savior, or come face-to-face with Christ after this life – as Judge only.

Third, Paul laid out the chronology of the biblical worldview for his listeners. The Creator of everything became a man, who defeated sin and death through his sacrificial atonement for this fallen world, and he will judge this world that he loves. In this regard, the Seven C's of History help make sense of the world we live in. The Seven C's are, in order, (1) Creation, (2) Curse, (3) Catastrophe, (4) Confusion, (5) Christ, (6) Church, and (7) Consummation, and each of these events from history tells the story of why Christ is the only true Savior of the world. The 'Case for a Creator' is a great way to begin sharing the faith with both seekers and skeptics, and it is how the Seven C's begin. However, as magnificent as the creation is, it is apparent that not all is well with the world. This is due to God's curse upon the world at the time of the rebellion, which directs us to the need for the cure: Christ. (Both the catastrophe of Noah's Flood and the scattering of people groups at Babel due to the confusion of languages solidify the accuracy of the beginning chapters of Genesis.) Christ's atoning for the sins of the world upon the cross is what has reconciled fallen man to the holy God who is revealed throughout Scripture. After Christ, the Church has kept the message of the Gospel alive and well in the world today. Finally, God will fashion a new heaven and a new earth one day, which we cannot even begin to imagine. This will be a world in which there is no pain or death, and everyone in his or her right mind will want to be a part of it – and hence the need for salvation through Christ.

---

[98] Galatians 3:28, NIV.

Table 2.1

The Seven C's of History

CREATION

CURSE

CATASTROPHE

CONFUSION

CHRIST

CHURCH

CONSUMMATION

The Seven C's are worlds apart from Greco-Roman evolutionism. That pagan philosophy maintains that there is no personal Creator, no fall of mankind and therefore no need for a redeemer or the Church that proclaims him, and no reason to believe the world will be upgraded for the better in the future. In Greco-Roman evolutionism, 'things just happen.' End of story.

Ultimately, the motivation for Paul in his Areopagus address was to see the pagan 'elites' come to a saving relationship with the one true God of the universe. This was not merely an academic exercise in comparative philosophy, pitting the biblical and Greek worldviews against each other. Granted, no one in the earliest decades of the Church could have engaged the Greek philosophers as effectively as Paul did. He possessed the ultimate combination of innate intelligence and religious education, yet ultimately it was the pastoral side of Paul – his heart for the lost – that shown the brightest in Athens. Paul's greatest apologetic tool was his love for people, not the impressive intellectual gifts that God had bestowed upon him. We all would do well to follow his shining example.

No matter what one believes about Paul's approach and content regarding this speech, there will always be some speculation involved, and we can never successfully argue a point that is lacking in scriptural evidence. In the end, the speech that we have is what we have, and Paul's Athenian address serves as the primary example of how to witness before an audience of well-educated non-Christians, both seekers and skeptics. Paul was the first great 'creation evangelist' to the world of evolutionism.

Evolutionism after the Apostles

In the second and third centuries, resulting from the Jewish diaspora and the spread of Christianity into Europe, Greco-Roman evolutionism was seriously challenged by the biblical creationists of the time. Theophilus of Antioch, an early Church Father, wrote in defense of biblical creationism in AD 170 while the Roman physician and skeptic Galen, in the very same year, wrote against it. Galen maintained that matter "was eternal, and his god was not above the laws of nature."[99] This type of pantheistic evolutionism eliminates the need for the one true God completely. The long-standing debate between biblical creationism and the idea of a chance origin of the universe continued from even more ancient times, and we are still debating it today.

Regarding Theophilus, it must be noted that his greatest work, *To Autolycus*, "contained an extensive treatment of creation and became a model for other fathers."[100] Although lengthy, it is worth including here the entire chapter on how the non-Christians of his time viewed God and creation:

> Some of the philosophers of the Porch [Stoics] say that there is no God at all; or, if there is, they say that he cares for none but himself…and others say that all things are produced without external agency, and that the world is uncreated, and that nature is eternal; and have dared to give out that there is no providence of God at all, but maintain that God is only each man's conscience [Epicureans]. And others again maintain that the spirit which pervades all things is God [monistic pantheists]. But Plato and those of his school acknowledge indeed that God is uncreated, and the Father and Maker of all things; but then they maintain that matter as well as God is uncreated, and aver that it is coeval with God. But if God is uncreated and matter is uncreated, God is no longer, according to the Platonists, the Creator of all things, nor, so far as their opinions hold, is the monarchy of God established.[101]

Neither the Stoics, the Epicureans, the monistic pantheists nor the Platonic philosophers would arrive at a correct interpretation of origins. But then again, how could they have? Without God's direct revelation on origins, they were doomed to follow mankind's speculation on the subject. Theophilus continues, this time focusing on the nature of the one true God of the universe:

> And further, as God, because He is uncreated, is also unalterable; so if matter, too, were uncreated, it also would be unalterable, and equal to God; for that which is created is mutable and alterable, but that which is uncreated is immutable and unalterable. And what great thing is it if God made the world out of existent materials? For even a human artist, when he gets material from someone, makes of it what he pleases. But the power of God is manifested in this, that out of things that are not He makes whatever He pleases; just as the bestowal of life and

---

[99] Lavallee.
[100] Ibid.
[101] Theophilus, *To Autolycus* (Book 2, Chapter 4). https://newadvent.org/fathers/02042.htm

motion is the prerogative of no other than God alone. For even man makes indeed an image, but reason and breath, or feeling, he cannot give to what he has made. But God has this property in excess of what man can do, in that He makes a work, endowed with reason, life, sensation. As, therefore, in all these respects God is more powerful than man, so also in this; that out of things that are not He creates and has created things that are, and whatever He pleases, as He pleases.[102]

In addition to Theophilus of Antioch, the great defenders of biblical creationism in the first few centuries of the Church included Justin Martyr, Irenaeus, Clement of Alexandria, Tertullian, Hippolytus of Rome, and Origen.[103] These early apologists often came out of a Gentile culture steeped in evolutionism: "The early fathers, like Justin, exchanged their evolutionary view of origins for the Biblical one. They believed the Bible and that God created all things out of nothing in the space of six days only a few thousand years ago, and once judged mankind with a worldwide flood."[104] Unfortunately, like Christian believers today they may not have been perfect concerning their theology: "One area where the Church Fathers went wrong was to accept uncritically the best Aristotelian science of their day; that is specifically geo-centrism and spontaneous generation."[105] The geocentric view can be forgiven, as this view – which held sway for centuries to come – does not necessarily lead to skepticism regarding the one true God of creation. Spontaneous generation, on the other hand, is a bit more difficult to understand in the light of Genesis 1-2. That likely demonstrates the power that the pagan philosophers held over natural philosophy (science) in the ancient world. Spontaneous generation would not be thoroughly refuted until the time of the Reformation, and despite the efforts of Redi and Pasteur it is still believed by naturalists today.

The early apologists had their hands full with the skeptics of their day, as these unbelievers were considered quite formidable. Besides Galen, who was possibly the greatest physician of the ancient world, the other great skeptic of Christianity in the second century was Celsus. Porphyry and Ammonius of Sakkas opposed biblical creationism in the following century. All were considered worthy opponents by their Christian opposition.

In the third century the apologist Lactantius confronted the evolutionary beliefs of the Greco-Roman philosophers. He noted that although they claimed many thousands of ages had passed since the world began, they were truly

---

[102] Ibid.

[103] It should be noted that both Clement of Alexandria and his student Origen held to an allegorical view of biblical interpretation, including the account of origins. Nonetheless, like the other great apologists of the early Church they battled against the Greco-Roman evolutionism of their day.

[104] Lavallee.

[105] Sibley, *Cracking the Darwin Code*, 28.

ignorant of origins and had greatly inflated the age of the earth: "Lactantius was also critical of pagans for pretending to know of the beginning, thinking that their stories (like the naturalism of today) could not be refuted. He insisted that the Bible exposed their falsity and the worldview they based on it."[106] There really is "nothing new under the sun."[107]

In the fourth century the apologist Arnobius continued to oppose the beliefs of the Epicureans, who were still influential at this time:

> Some men deny the existence of any Divine power. Others inquire daily as to whether or not one exists. Still others would construct the whole fabric of the universe by chance accidents and by random collision, fashioning it by the movement of atoms of different shapes.[108]

Clearly the teachings of Democritus, Epicurus, and Lucretius continued to present a challenge to Christ's followers, even centuries after the Church began. Yet so many believers today mistakenly assume that the origins debate only goes back to the time of Darwin.

The key defenders of biblical creationism in the last few centuries prior to the Middle Ages included Ephrem the Syrian, Basil the Great, Jerome, Cyril of Jerusalem, Ambrose of Milan, and Augustine of Hippo. Many of the great theologian-apologists of the early Church era were staunch defenders of a literal six-day creation period as outlined in Genesis 1, a young earth, and a global Flood in the days of Noah.[109]

Augustine and the Language of Genesis

In the time of the early Church there was another point being debated among Christian thinkers: Was all the matter that makes up the universe brought into existence 'in the blink of an eye' at the initial moment of creation on Day One, or did God create new matter on a day-by-day basis throughout the creation week? For some theologians, God created all the matter in an instant, and then fashioned that matter into stars, planets, nebula, mountains, oceans, animals,

---

[106] David Green, "The Long Story of Long Ages." https://answersingenesis.org/genesis/the-long-story-of-long-ages/

[107] Ecclesiastes 1:9, NIV.

[108] Arnobius, *Against the Heathen* (Book 1, Chapter 31). https://newadvent.org/fathers/06311.htm

[109] The major exceptions to a straightforward interpretation of biblical creationism included the Alexandrian theologians Clement and Origen, and possibly Augustine of Hippo. Clement and Origen held to an allegorical (symbolic or figurative) approach to biblical interpretation, while Augustine's approach to interpreting Scripture is debatable. He seemed to favor an instantaneous creation of all matter at the initial moment of the first day, followed by a fashioning of that material into everything we see over the remainder of the six-day creation week. That is, in my opinion, not even close to being heretical. He did not, however, plainly support creation over vast eons of time, although some old earth creationists and theistic evolutionists believe he did.

people, and so forth over the course of six days. For other theologians, God created matter every time he brought something new into existence during the creation week. There is nothing heretical with either view, but it was a point of debate among some Christian thinkers in the early Church. No less than Augustine (AD 354-430), who is almost universally considered to be the greatest theologian after the apostles, held to the first approach.[110]

AUGUSTINE

The Church historian Benno Zuiddam comments on Augustine's view: "Augustine was not vague about the age of the earth, the historicity of Adam and Eve as our first ancestors, or the events in the Garden of Eden and the worldwide flood later in Genesis. However, his doctrine of creation was complex. All matter, according to him, was created on the first day…"[111] For Augustine, God created all the matter of the universe 'in the blink of an eye.' Then, during the six days of creation God fashioned everything out of that matter.

Some old earth creationists claim that Augustine held to this view because of the Hebrew words used in Genesis. This involves the words *bara* ('to create') and *asah* ('to make'). Terry Mortenson, a recent creationist who earned his doctorate in the history of geology, comments on the language of Genesis creation:

> They [old earth creationists] sometimes try to defend the acceptance of millions of years by saying that *bara* refers to supernatural creation *ex nihilo* (Latin for 'out of nothing') but that *asah* means to make out of pre-existing material and therefore allows for creation over a long period of time. Such people say that the only supernatural creation events were in relation to the heavens and the earth (Genesis 1:1), sea creatures and birds (1:21), and Adam and Eve (1:27). Since *asah* is used for all other

---

[110] Although no one can exceed or even equal the importance of the apostles – other than Jesus, of course – Augustine has done more to impact theology than any other Christian thinker since his time.

[111] Benno Zuiddam, "Augustine: Young Earth Creationist." https://creation.com/augustine-young-earth-creationist

creative acts in Genesis 1, those acts could have been creative processes over the course of millions of years.[112]

Some creationists and biblical scholars claim that when the word *bara* is used, it denotes an event that was accomplished quickly, perhaps even 'in the blink of an eye' as noted above, while the word *asah* indicates a process that could take a much longer time, maybe even millions of years or more. Not all scholars agree with this notion. Some scholars argue that a *bara* event could take lots of time, while an *asah* process could be accomplished quickly. The old earth creationist Jon Greene points out the main distinction between these words:

> The Hebrew word for 'created' (*bara*) means to create out of nothing. It is a completed verb form, meaning only that the creation was accomplished at some point in the past…In contrast to the verb 'create' (*bara*), the verb *asah* expresses making something from pre-existing material, not *ex nihilo* creation of Genesis 1:1.[113]

On the other hand, Mortenson contends that these two words may be used interchangeably:

> *Bara* (create) and *asah* (make) are used interchangeably in the Bible in reference to the creation of the following: the sun, the moon, the stars, sea creatures, trees, rivers, man, the heavens, and the earth. In several verses they are even used together to describe the same event.[114]

Like most other things associated with the origins debate, young earth and old earth creationists dispute each other over the Hebrew language used in the Genesis creation account. I strongly suspect that Moses, writing under the inspiration of God's Spirit, did not randomly choose the words *bara* and *asah*. Instead, every word in both the Old and New Testaments was surely chosen with intention. There is almost certainly a reason why *bara* appears where it does, while *asah* is used in other places. These words may very well be used to describe the same events, but there must be some reason for their exact placement. Peter reminds us that God is the ultimate author of Scripture: "For prophecy never had its origin in the human will, but prophets, though human, spoke from God as they were carried along by the Holy Spirit."[115] I am confident that God inspired the biblical authors to choose their words carefully.

---

[112] Terry Mortenson, "Understanding Genesis 1 Hebrew: Create (*bara*) & Make (*asah*)." https://answersingenesis.org/genesis/did-god-create-bara-or-make-asah-in-genesis-1/

[113] Greene.

[114] Mortenson.

[115] 2 Peter 1:21, NIV.

Let us return to the idea that Augustine's view on creation was formed largely through a strict adherence to the biblical languages. To say that Augustine was a great theologian would be an understatement. Almost all theologians consider him to be the greatest post-apostolic theologian of the first 1,000 years of the Church era, and maybe the greatest theologian since his time. However, despite his great intellect it is generally acknowledged that Augustine was not especially strong in Greek, nor was he exactly enamored with Hebrew:

> Jerome took the trouble to learn Hebrew, which Augustine thought was unnecessary, since he believed that God had inspired the Greek translation known as the Septuagint. That made the Hebrew original obsolete, in Augustine's eyes, and most of the church at that time agreed with him…Unfortunately, Augustine's Greek was not very good either, and he struggled with the biblical text.[116]

Despite this lack of proficiency in, or even need for, the original biblical languages, we must keep in mind that Augustine was a true scholar in every other way. Therefore, he knew where to find answers. If he lacked certainty about the Hebrew words chosen for Genesis 1-2, he could have consulted the experts – there was an abundance of Greek-speaking theologians in his time, who were able to tackle the Septuagint (Greek Old Testament) with no problem. He could have consulted them. I am confident that he arrived at his theological conclusions based upon solid reasons, and that includes the original language of Genesis. Augustine did nothing 'willy-nilly.'

I have no reason to doubt Augustine's idea of the 'raw material' being created first, followed by God fashioning that raw material into everything that exists over the remainder of the creation week. Perhaps the important point is this: Augustine's ideas were a far cry from both the Gnosticism and naturalism that existed in his day. Zuiddam notes that Augustine was, despite the claim of some old earth creationists and theistic evolutionists, a recent creationist:

> But, all speculations set aside, Augustine did not teach a process of one kind changing into another…In Augustine's mind, God would have created all matter as well as the seminal ideas in the blink of an eye. The material expression of those ideas followed later. We have to combine his instant creation theory with his literal reading of other events in Genesis. Adding his belief that the world is about 8,000 years old makes it extremely

---

[116] Gerald Bray, "Augustine's Key."
https://christianitytoday.com/history/issues/issue-80/augustines-key.html

hard to call on him to support Darwinian evolution of any kind or deep time.[117]

Augustine clearly denied the evolutionary model of origins, which was prevalent among many Gentiles. You may or may not find his view on creation strange. I see nothing strange, or even heretical, about it. Compared to the ideas of Greco-Roman evolutionism, his ideas on origins were clearly in accordance with Scripture.

'There is Nothing New Under the Sun'

As we have seen, evolutionism was prevalent in the cultures of the early Church era. Yet even at that time it was already an ancient idea, having been part of the beliefs of Babylon, Egypt, India, Greece, and Rome. The Church in its first few centuries of existence fought against evolutionism, and we are still fighting against this belief today, in the modern guise of Darwinism. Unfortunately, the battle now unfolds not only within the culture, but has moved into the Church as well. Many well-intentioned, but deceived, believers have attempted to integrate evolutionary ideas into the early pages of Genesis. As Solomon wrote three thousand years ago, "there is nothing new under the sun."[118]

In the next chapter we will find ourselves traversing many centuries, beginning with the Middle Ages and ending with the Enlightenment or 'Age of Reason.' (This is a lengthy period, beginning in the middle of the fifth century and ending around the early 1800's.) Concerning the origins debate, a lot happened in this time. The Church continued to uphold the doctrines of God, creation, and biblical reliability in the Middle Ages, but the Renaissance and especially the Enlightenment eras challenged the Church's influence in every way. The Church was no longer the only real authority in life's matters: Man was once again becoming the 'measure of all things.' Humanism, which had thrived in the Greco-Roman world, began to make a comeback in the Renaissance – and it took off 'big time' in the Enlightenment. The sacred and the secular were once again locked in mortal combat.

Protagoras of Abdera, a Greek philosopher who lived in the fifth century BC, wrote that, "Man is the measure of all things: of the things which are, that they are, and of the things which are not, that they are not."[119] It took some time – centuries, in fact – but eventually man did become 'the measure of all things' regarding existential matters. For many people, God was on the way out – and man was there to take his place.

---

[117] Zuiddam, "Augustine: Young Earth Creationist."
[118] Ecclesiastes 1:9, NIV.
[119] Quoted by Plato in *Theaetetus*.

CHAPTER THREE

Origins: Middle Ages to the Enlightenment

The battle between Greco-Roman evolutionism and biblical creationism continued throughout the Middle Ages, Renaissance, Reformation, and Enlightenment eras. The Church in the Middle Ages strongly upheld the doctrine of biblical creationism, but by the time of the Enlightenment doubt and disbelief in the early chapters of Genesis became the norm for many scholars of the day. Unfortunately, their secular influence was powerful in Western culture, and the Church is still reeling from this time of great skepticism. Sadly, from the time of Darwin to the present evolutionism has become even more pronounced throughout Western culture.

Yet most people today are surprised to learn that it was the Renaissance and the Enlightenment eras that really paved the way for Charles Darwin. Without those two periods of history, there probably would not have been a Charles Darwin's *Origin of Species*.[120] Reformation thinking – with its reliance upon Genesis creationism – may have helped stem the tide of pagan evolutionism, but in the end fallen man was ready to become 'the measure of all things.'

Evolutionism in the Middle Ages

The Middle Ages began in the mid-to-late 400's, when the Roman Empire fell to the surrounding nations of the European front, and they ended in 1400 with the birth of the Renaissance.[121] Some historians and scholars refer to the Middle Ages as the 'Dark Ages,' but this is a derogatory term that was actually begun during the Renaissance era by people who felt intellectually superior to those who immediately preceded them. The Middle Ages was not a period marked by a lack of knowledge and understanding, however. The creationist Jonathan Sarfati notes:

> While this period used to be called the 'Dark Ages,' responsible historians recognize that it was far from dark. Rather, it was a period of great scientific advances, stemming from the logical thought patterns of the medieval Scholastic philosophers of the Church, and the extensive inventiveness and mechanical ingenuity developed in the monasteries. Small wonder that this

---

[120] Charles Darwin likely would not have been so keen about developing his philosophy of macroevolutionary development had there been no Renaissance or Enlightenment eras. Western culture was primed for his beliefs because of this secularized time in history.

[121] Likewise, some have claimed that the Middle Ages began with the death of Augustine in AD 430, whereas others point out that Rome officially fell in AD 476. Regardless of the exact date, Augustine is often considered to be the last of the Church Fathers and the first of the Medieval theologians, bridging the gap between the ancient world and the Middle Ages.

period saw the development of water and wind power, spectacles, magnificent architecture, the blast furnace, and the stirrup.[122]

Far from being the 'Dark Ages,' the Middle Ages saw many important scientific discoveries that helped lay the foundation for modern science. Although it is widely believed by many today that the Church stifled science during this time in history, this was simply not the case. The Church never proclaimed a flat earth, never banned human dissection, and never burned anyone at the stake for advancing scientific discoveries.[123] The Church was the sole bastion of higher learning during this time, keeping alive both biblical truth and natural philosophy (science) in an otherwise illiterate world. "It should not be forgotten that, from the fifth to the ninth centuries, Christian missionaries were spreading the light of the gospel from the British Isles to Russia to those in the clutches of the spiritual darkness of paganism."[124] This paganism generally included evolutionary ideas such as the spontaneous generation of life from non-living matter and some form of naturalistic progression from lower-to-higher life forms. Those who were caught up in the errors of paganism were instead given the truth of the Gospel, including biblical creationism, by those who were willing to carry out the Great Commission.

It should also be mentioned that the university system of education began in the Middle Ages. The University of Bologna (Italy) is the oldest institution of higher learning in Europe, having begun in 1088. Oxford University (England) is nearly as old, however: There is evidence that teaching took place there in 1096, although some claim that this institution dates back even earlier. The University of Salamanca (Spain) was founded in 1134, while the University of Paris (France) began shortly after that. There were several other universities that began in the Middle Ages as well, including Cambridge University in England (1209), the University of Padua in Italy, (1222), and Charles University in Prague, Czechoslovakia (1348). An even earlier university pre-dated all these: Al-Azhar University, in Cairo, Egypt was founded as early as AD 970. It originally began as an educational institution for all levels of learning, from youth through the highest degrees obtainable. Although it focused on Islamic

---

[122] Jonathan Sarfati, "The Biblical Roots of Modern Science." https://creation.com/biblical-roots-of-modern-science

[123] A little later in history the scientist Galileo Galilei (1564-1642) would run counter to the Church when he proclaimed the earth is not the center of the solar system. Nonetheless, he was not burned at the stake for attempting to advance this idea, but rather was confined to house arrest. Still harsh, in my opinion, but not gravely so. On the other hand, the Dominican friar Giordano Bruno (1548-1600) was burned at the stake, but for theological disagreements with the Roman Catholic Church – not scientific claims. This included a denial of several core doctrines such as eternal damnation, the Trinity, the divinity of Christ, the virginity of Mary, and transubstantiation. In other words, all the major beliefs of Roman Catholicism. Bruno was also believed by some to have dabbled in the occult as well. This execution should never have happened, and it would not today, but it was a vastly different world in the sixteenth century.

[124] Herbert, 31-32.

studies in the beginning, it now provides several programs of study. If the university system began in the Middle Ages, then the people of that time were certainly not the unenlightened imbeciles that they have been made out to be by later generations. After all, by-and-large the people of this time did choose the correct side in the theism-atheism debate – which is something that many of our most educated 'elites' cannot seem to do today.

In AD 632 a new competitor to Christianity was born. Muhammad Ibn Abdullah, a wealthy merchant from Arabia, founded the religion of Islam. The Qur'an (Islam's Holy Book) holds the same basic belief in the Creator that the Bible teaches, although in a very general sense only. (There are many differences in both God's character and the details of creation between the Bible and the Qur'an, which is beyond the scope of this study.) Islam replaced the polytheistic evolutionism of Arabia with creationism by one sole Creator-God.[125] Approximately half of the world's population today is either Christian or Muslim, which means that half of the people in the world hold to a belief in one supreme Creator-God.[126]

Once they were rediscovered, the Muslim scholars of the Middle Ages embraced the Greco-Roman philosophers. They were especially impressed with the writings of Plato and Aristotle: "They considered the works of Plato and Aristotle to be equal to the Judeo-Christian Scripture in providing inspiration for reading the Qur'an."[127] Unfortunately, this influence made its way into the Church in short time. "From the Middle Ages Christian thinkers found much attraction in Plato's work despite his affinity for the existence of a pantheistic world soul, or *anima mundi*, and rejection of creation out of nothing."[128] The concept of creation out of nothing (creation *ex nihilo*) is a foundational biblical doctrine, so this was truly an unfortunate influence for the Church in the Middle Ages. (The very first verse of the Bible screams 'creation out of nothing.') These ideas of Plato held much in common with Hinduism, which also maintain a belief in an eternal universe and a pantheistic world soul. As mentioned in chapter one, it was likely Pythagoras who originally introduced the ideas of Hinduism to the Greek isles.

Let us take the 'off ramp' for a moment, for a quick-but-beneficial side study. Another commonality between Hinduism and Greek thought is the idea that the physical creation is merely an illusion. Only the spiritual aspect of creation is believed to be real, while the physical universe is considered illusory. In Hinduism, this idea is referred to as *maya*, a Sanskrit word meaning 'illusion.' The Gnostics, who rose to prominence in the late first century, further developed this idea alongside the new religion of Christianity. The Gnostics were noted for

---

[125] That is not to say that the God of the Bible and the God of the Qur'an are one and the same, as there are clearly marked differences between the two.

[126] Once again, in a very general sense.

[127] Sibley, *Cracking the Darwin Code*, 32.

[128] Ibid.

their distaste of all things physical, including sexual activity. (This is interesting, in that the Gnostics still had children like everyone else!) The Bible, on the other hand, describes both a spiritual dimension and a physical creation. We live in a material world that is real – its creation is plainly described in Genesis 1 – yet at the same time we are surrounded by a spiritual dimension, which we will someday inhabit. This demonstrates the strong Eastern influence upon Greek thought.

The key defenders of biblical creationism in the Middle Ages were extremely devoted to the Church. Included in this group are Isidore of Seville, John of Damascus, Albert the Great and his star pupil, Thomas Aquinas.

Thomas Aquinas: Elevating Reason

Although Thomas Aquinas (1225-1274) was a devout Christian theologian, trained in Augustinian theology, he was undoubtedly responsible for introducing the teachings of Aristotle to the Church. He was the first to refer to Aristotle as 'The Philosopher,' almost as if there were no other great thinker's worth noting from the Greco-Roman world. This non-Christian influence from Aquinas later inflamed Martin Luther in the time of the Reformation, for Luther correctly noted that no pagan teachings could ever be allowed to supplant or even attempt to 'compliment' biblical doctrine. For Luther – and for us today – it is *Sola Scriptura*, or 'Scripture alone,' with no need for non-Christian philosophy of any sort, no matter how well-intentioned it may be.[129]

THOMAS AQUINAS

Aquinas argued that, even apart from Scripture, all people can know of God's existence and all-encompassing power simply by observing nature. This is a belief that countless Christians agree with wholeheartedly. He held to this idea because of his view of the fall of man. Whereas Augustine was adamant that the

---

[129] The debate between the role of philosophy in the Church is still ongoing. In the early Church, Justin Martyr fully supported the use of Greco-Roman philosophy, provided it did not go against key biblical doctrines, while Tertullian vehemently opposed it. Tertullian noted in chapter seven of *The Prescription Against Heretics*, "What indeed has Athens to do with Jerusalem? What concord is there between [Plato's] Academy and the Church?" For him, the answer was no concord whatsoever. Scripture and philosophy should never be mixed, as this always results in a 'watering down' of biblical truth.

fall of man negatively affected every aspect of humanity, including the intellect, Aquinas believed that the intellect was left unaffected by the fall. For Augustine God was 100% responsible for bringing fallen man to the truth of salvation, whereas in Aquinas' thinking man could, at least in some small part, reason his way to God through an examination of natural philosophy (science).

Aquinas developed his 'five ways' as proof for God's existence. These are the proof from motion, the proof from efficient cause, the proof from necessary being, the proof from degrees of perfection, and the proof from order in the universe. In the proof from motion, Aquinas taught that whatever is moved must be moved by some prior actuality, but without an infinite regress. Therefore, a 'Prime Mover' or 'First Cause' must exist. God is the eternal, uncaused Prime Mover who set everything in motion. The proof from efficient cause relies on the law of cause-and-effect. (See chapter five for a more extensive explanation of this evidence for God, which is commonly referred to today as the cosmological argument.) The law of cause-and-effect states that everything which has a beginning has a cause, and the cause always transcends the effect. Since the universe had a beginning (Genesis 1:1), the universe had a cause. What, besides God, could transcend the universe? For Aquinas, the answer was obvious, as it should be for all of us (Romans 1:20). The proof from necessary being is, with some modification by Aquinas, the same as the ontological argument first proposed by Anselm of Canterbury a few centuries earlier. The ontological argument for God's existence maintains that God is the greatest being that can be imagined. Since you can imagine the God of the Bible (albeit never perfectly), he therefore exists. This is not my favorite argument for God's existence, but it works well for some theologians. The proof from degrees of perfection readily lends itself to the moral argument for God's existence. Borrowing heavily from Augustine for this proof, Aquinas argued that it is impossible to have a relative comparison of anything without an absolute with which it can be compared, and that absolute is the perfection found only in God. (Interestingly, Augustine largely borrowed this idea from Plato. Aquinas was not the only great theologian who was guilty of utilizing Greco-Roman philosophy in his apologetic.) Finally, the proof from order in the universe is the same as the design argument for God's existence. In short, an intelligent design demands an intelligent Designer.

Table 3.1

Thomas Aquinas' 'Five Ways'

PROOF FROM MOTION
(PRIME MOVER OR FIRST CAUSE OF EVERYTHING)

PROOF FROM EFFICIENT CAUSE
(COSMOLOGICAL ARGUMENT FOR GOD)

PROOF FROM NECESSARY BEING
(ONTOLOGICAL ARGUMENT FOR GOD)

PROOF FROM DEGREES OF PERFECTION
(MORAL ARGUMENT FOR GOD)

PROOF FROM ORDER IN THE UNIVERSE
(DESIGN ARGUMENT FOR GOD)

Aquinas' five-point 'Case for a Creator' points us to the existence of the all-powerful God of the universe, and an intellectually honest seeker of truth would do well to consider these lines of evidence from Aquinas. However, openness to following the evidence, no matter where it might lead, is a requirement in the pursuit of truth. For many people who are skeptical of God and spiritual matters, this may be more than they are willing to do (1 Corinthians 2:14). Although Aquinas was quick to point out that recognizing God through nature did not save anyone, it was enough to at least point them in the right direction until God further illumined their darkened hearts and minds, leading them to salvation in Christ (Romans 10:9).

Although Aquinas 'takes a beating' by many in the Church, it must be noted that he was, first and foremost, devoted to the scriptures. Regarding creation, he was adamant that the universe had a beginning, despite the commonly held Greco-Roman belief in an eternal universe. So even though he greatly entertained the writings of the pre-Christian philosophers, he did not buy into their non-scriptural ideas in every way, as some suppose.

The debate concerning the impact of mankind's sin on our intellect is still raging today. Some Christians are quick to note that it is only with the guidance of Scripture that we can truly and completely know that God exists, whereas others readily note Paul's reference to natural theology in Romans 1:20. In the natural theology debate, there are valid arguments for both sides.[130]

All in all, during the Middle Ages the biblical worldview was maintained from the era of the early Church, including special creationism as outlined in Genesis 1-2. However, this would soon be challenged during the time of the Renaissance, when man's reasoning abilities became the supreme basis for knowledge, as opposed to a reliance upon the clear teachings of Scripture.

Evolutionism in the Renaissance & Reformation

Following the Middle Ages, the Renaissance was the next great period in history, beginning in approximately 1400. It was a time marked by a renewed interest in learning, not just for the privileged but for the masses as well. Unfortunately, this learning often took place totally apart from Scripture. "Mankind now became the center and measure of all things. In the ultimate questions of life, human reason was rapidly supplanting the Scriptures as the final authority."[131] That trend began with Aquinas and others before him, and only intensified in the centuries to follow. During the Renaissance there were still many who held to the biblical concept of creationism, but the newfound reliance upon the Greco-Roman philosophers challenged the early chapters of Genesis.[132] Evolutionism in the Renaissance era was on the rise once again.

Later in the Renaissance, around 1517, the Reformation began. Renaissance ideas, which borrowed heavily from the Greeks and Romans, primarily affected southern Europe. On the other hand, the Reformation – with its' clear-cut emphasis upon *Sola Scriptura* or 'Scripture alone' – was seen mostly in northern Europe. Therefore, evolutionism at this time was more likely to receive an audience in southern Europe as opposed to the northern nations, which became even more reliant upon biblical creationism during this time.

When it came to reading Scripture, the general Renaissance approach involved reading symbolism into the text, as opposed to a straightforward reading. There is symbolism and allegory found within the pages of the Bible, to be sure, but Renaissance theologians often made the mistake of inserting symbolism into areas where there should be none. Fortunately, the Reformation changed that approach and people began to read the historical portions of Scripture as straightforward history once again:

> For much of Church history Christianity accommodated [ancient pagan] texts by seemingly Christianizing them, but with the Reformation came a greater commitment to the literal text of

---

[130] The natural theology debate is beyond the scope of this book, but all Christians should be encouraged to study the issue. Scripture that seems to support natural theology includes Psalm 19:1 and Romans 1:20, while Proverbs 3:5 indicates that we cannot fully recognize the Creator apart from his direct revelation in Scripture.

[131] Herbert, 40.

[132] It may be accurate to say that, for the masses, biblical creationism took precedence while the scholarly elite strongly entertained non-biblical ideas such as deism and naturalism.

the Bible with the symbolic Renaissance interpretations downplayed. The Reformation, and the translation and printing of Bibles, enabled the corresponding scientific reformation in large part. Science developed because it opened up a way of reading the created order literally, and encouraged respect and equality in education and the pursuit of knowledge. As Peter Harrison has noted, a more literal reading of Scripture encouraged a more literal reading of nature, which led to developments in operational science.[133]

The next time someone tells you that a literal reading of Genesis 1-2 is 'not scientific,' you may want to inform him or her about the origins of early modern science. The founders of many of the branches of science were Bible-believing Christians who took Genesis 1-2 very seriously. They could plainly see God through creation.

Key defenders of biblical creationism in the Renaissance and Reformation eras included Martin Luther, Philip Melancthon, and John Calvin, the greatest theologian of the Reformation period. "Calvin revered the early Church Fathers but he had difficulty accepting Augustine's allegorical approach to Genesis. Rather, he admired Chrysostom for his straightforward interpretation of the Bible."[134] Without exception, the Reformation leaders were presuppositional in their apologetic approach, meaning that they argued from the Bible, not to the Bible as evidential apologists do. Led by their commitment to *Sola Scriptura*, they used logical evidence from science and nature far less than many apologists do today, instead relying on the truth of divine revelation without the need to argue for biblical doctrines.[135]

Rene Descartes: Further Elevating Reason

At the tail-end of the Renaissance a man was born who would change the face of the origins debate for the next several centuries. Rene Descartes (1596-1650), the "father of modern philosophy,"[136] was the founder of philosophical rationalism. Rationalism places reason well above faith, although it must be noted that not all rationalists rule out faith completely. (Descartes, as a devout Roman Catholic, certainly did not completely rule out faith in his life.) The elevation of reason began with Thomas, although he was undoubtedly a man who exercised great faith as well. Descartes merely amplified this rationalistic approach. Descartes was raised in the Roman Catholic Church, and we can say with certainty that he never doubted God's existence. His focus on rationalism

---

[133] Sibley, *Cracking the Darwin Code*, 34.
[134] Herbert, 50.
[135] This is not to say that they completely avoided logic-based evidence. Some, like Calvin, were more likely than others to refer to the logical evidence behind creationism.
[136] Sproul, *The Consequences of Ideas*, 82.

did, however, open the door for future thinkers who would question the reality of God.

RENE DESCARTES

Of course, both faith and reason play a role in the life of the believer. Paul clearly tells us that the righteous live by faith (Romans 1:17; Galatians 3:11), and we are supposed to trust God and not lean on our own understanding (Proverbs 3:5). It seems that we are supposed to trust God through faith, regardless of whether his words always make perfect sense to us.

On the other hand, God tells us to use the intellectual gift of reason (Isaiah 1:18). We are to have good reasons for what we believe about God and spiritual matters, and we must always be ready to share those reasons with others (1 Peter 3:15). Therefore, we must attempt to show unbelievers that our belief in God, Christ, and the divine authority of Scripture is reasonable, justified, and can be logically defended.

So, is it faith or reason that is the basis for our belief in God and the Christian faith? Are we supposed to rely solely upon our intellect, using reason to make sense of biblical faith while rejecting those things which seem to be irrational? Or should we instead accept the teachings of Scripture without any regard to logic and reasoning? As based upon Scripture in its entirety, clearly both faith and reason are meant to work together. One should not dominate the other, as they are intended to inform us in complimentary ways.

Therefore, reason cannot be elevated above faith. Apologetically, there are at least three reasons for this. First, the author of Hebrews makes it clear that without faith, it is impossible to please God (Hebrews 11:6). Reason alone cannot produce faith. Reason supports faith, and reason can point to faith in Christ as a logical conclusion, but only God's Spirit produces faith itself.

Second, there are some truths that are known only through God's revelation in the written word. Perhaps the best example of this is God's triune nature. God's general revelation expressed through creation can point us in the direction of the Creator, but the obvious design that we observe in this world cannot point us specifically to God's triune nature. For that we need God's self-revelation in Scripture. The six-day creation plan is another example of something we know about only because God has revealed it to us.[137] Looking at nature, we can see the 'fingerprints' of the Creator everywhere we look, but we would not know that God created everything in the span of six days just by observing nature.

Finally, rationalism fails to account for the difference between the rational-thinking mind and the emotional-volitional will. There are many examples in our modern culture of people who have acknowledged the evidence for God in general and even Christianity in particular, but nonetheless they chose not to act positively upon that evidence and instead remained comfortably settled in their non-belief. Just because there may be good evidence's for believing in something – and Christianity is a great example of a worldview with many powerful evidence's in its favor – that does not mean that everyone will choose to accept those evidence's and embrace the Christian life. For that, we require faith, which is a gift from God (1 Corinthians 2:13-14).

Descartes was almost certainly very well-meaning, but nonetheless he played a significant role in further elevating reason above faith. Not surprisingly, some philosophers and theologians that came after him elevated reason even to the point of suppressing faith. Examples include the Jewish philosopher Baruch Spinoza (1632-1677) and the deists of Europe, who continued to elevate reason above faith but in a far more skeptical direction. They would trade the one true God for a god who is either an impersonal force (pantheism), or a god who creates but has no interest in further involving himself in the affairs of man (deism). For these people, reason was elevated so far above faith that faith became unnecessary, or even counterproductive to reason in some cases. Sadly, Western society has never recovered from this attempt to free humanity from the life of faith.

Evolutionism in the Enlightenment

Around 1600, mankind entered what has commonly been termed the Enlightenment or 'Age of Reason.' Continuing to push the teachings of Scripture to the sidelines, rebellious man began to shake himself free from what he saw as the confines of the Church. Fortunately, many people remained steadfast to the Word of God, resulting in a worldview battle that has not abated to this day.

In 1650 James Ussher, the Archbishop of Armagh, Ireland published a history of the world from the creation to AD 70, the year that the Jewish Temple was destroyed by the Romans. The Ussher Calendar proposed a creation date of 4004 BC, which was widely accepted at the time – and still is today by many creationists. Although Ussher may have 'pushed the envelope' a bit by being too precise with his chronology, even offering the time of day that creation began, his estimation for the age of the earth was extremely well-researched. In his time,

---

[137] Without the six-day creation plan outlined in Genesis 1, we would still know that God exists (Romans 1:20), but we would be no better off than an Intelligent Design (ID) proponent who simply believes that the creation was intentional in a very broad sense. God has specifically revealed to us how he created, in the span of six days.

however, there were many who began to push for an ancient earth, millions if not billions of years old. The battle over the age of the earth, which was anything but a recent development, ensued with great fervor from Ussher's time to the present.

Besides the age of the earth, the Noahic Flood was another major issue in the time of the Enlightenment. Concerning the Great Flood, many differing viewpoints were offered by the scholars of the day. Many influential thinkers, especially among the French skeptics, opposed the idea of the Flood altogether while others argued convincingly for its historicity in various watered-down forms. (The idea of a local or regional flood was proposed by many skeptics.) In 1696 the English theologian William Whiston became the first to propose that the Noahic Flood was caused by the water in the tail of a comet that had struck the earth. Even today many have suggested that a heavenly body was responsible for initiating the Flood – at God's command, of course.

Besides Ussher and Whiston, other key defenders of biblical creationism in this period included John Lightfoot, John Ray, and Isaac Newton.[138] The Westminster Confession of Faith, written in 1646, declared the truth of God's creation out of nothing. It thoroughly refuted the Greco-Roman evolutionism that was being revived by many of the skeptics of the day.

The Scottish scientist James Hutton, considered by many to be the first modern geologist, proposed the concept of uniformitarianism during this time. (However, it should be noted that the French skeptic Benoit de Maillet, to be mentioned shortly, conceived of the idea even earlier.) Uniformitarianism is the belief that geologic processes have been extremely slow and constant throughout time. The motto of uniformitarianism is, "The present is the key to the past." In other words, the slow, gradual processes of nature that we observe today are wholly responsible for shaping the earth throughout its great history. Uniformitarians recognize that cataclysmic events have taken place on occasion, but ultimately it is the slow, gradual process of change over millions of years that created the geologic features that we see today. Not surprisingly, Hutton proposed a great age for the earth, while other skeptics maintained a belief in a world without a beginning. Many, if not most, creationists considered the earth to be quite young up until this time, as based in large part upon the genealogy from Adam to Jesus that is recorded throughout the Bible. Hutton's arguments contributed greatly to the formation of old earth thinking, which would be later expanded upon by Charles Lyell, a contemporary of – and an inspiration for – Charles Darwin.

---

[138] Isaac Newton remains very controversial as a theologian and apologist, however. It is widely-held that he was a non-Trinitarian, which was considered heretical in his day – and not always well-received today, for that matter! Much of his theological writings, which are believed to have equaled if not surpassed his scientific writings, focused largely on biblical chronology and the decoding of prophecy. Unfortunately, like many in his day Newton believed that reason was a primary means of knowing God. Reason, although necessary for correctly interpreting Scripture, must not relegate revelation and faith to the background.

Around this same time, the Scottish philosopher and skeptic David Hume argued against intelligent design in nature. He taught that nature appeared to be intentionally designed but was, in fact, attributed to chance, random processes in nature which merely gave the appearance of design. This was nothing new, as it is identical to the teaching of Epicurus three centuries before Christ. The English philosopher William Paley responded to Hume by refining the ancient design argument first proposed by the Greek philosophers and Cicero in his 1802 publication *Natural Theology*. His watchmaker analogy is still used today as an argument for divine design. Paley noted that if you were to find a watch on the ground, you would not assume that it randomly 'evolved' from the elements found in the earth, but rather it shows every indication of having been intentionally designed by someone. Just as a watch requires a watchmaker, an intelligent design requires an intelligent Designer. When it comes to nature, intelligent design is all around us, demonstrating the existence of the intelligent Designer.

Hume is interesting in that most scholars today–both skeptics and believers–consider him to have been an atheist. However, he seemed to affirm God's existence in some of his writings; it was God's nature that seemed to be unclear for him. Hume may have been a deist, which certainly fits well with the time he lived in. Regarding design in nature, Hume claimed that the design argument does not prove one specific God, nor does it prove that God is infinite. That may be true, but the point of the theistic arguments[139] is that they point people in the direction of the one Creator-God, while other, more specific Christian arguments take you the rest of the way to the one true God revealed in Scripture. This was my approach in *Worldviews in Collision*: I first laid out the case for theism (God), then followed that with the 'Case for Christ' (Christian God) in the second half of the book. This is essentially the classical apologetic or 'two-step' approach to proclaiming and defending the faith. For those who want to explore this apologetic method in more detail, I heartily recommend my first book. (A shameless advertisement if ever there was one!) If you are interested in further exploring Hume's skepticism toward design in nature, I highly recommend Shaun Doyle's "David Hume and Divine Design," which may be found at Creation Ministries International (creation.com).

Francis Bacon and the Scientific Method

The English statesman, lawyer, and philosopher Francis Bacon (1561-1626) is credited with developing the scientific method. He correctly noted that we must not rely upon the ideas of Aristotle or any other philosopher when explaining the origin and function of natural things, but he also maintained that we must not rely solely upon "Genesis 1, Job, or any other part of the Bible"[140]

---

[139] The classic theistic arguments are the cosmological, teleological, ontological, and moral arguments. I describe these in depth in *Worldviews in Collision*.

[140] Jonathan Sarfati & Carl Wieland, "Culture Wars: Bacon vs Ham (Part 1)."

as well. Although this idea will work for purely experimental science (operational science), it does not work for determining origins (historical science). For historical science, we must rely upon one of two things: Speculation, or the revelation given to us by One who was there. This explains why God's revealed words on creation are so important to Bible-believing Christians. Some creationists claim that Bacon not only opposed "building science on Greek philosophies, but he also rejected the Bible itself as a basis of scientific knowledge."[141] I am not so sure about that, however. He did oppose both Greco-Roman philosophy and Scripture in experimental science, but not necessarily concerning historical science.

FRANCIS BACON

Like Thomas Aquinas, Bacon sometimes 'takes a beating' in the Church. In the process of removing one bad influence from science – erroneous pagan writings about origins – some scholars suggest that Bacon also claimed we should not fully rely upon God's revealed words on origins, either. But we know that God's revelation is necessary to be truly informed about creation. Was Bacon really saying that Scripture has no place concerning origins?

First and foremost, it must be noted that Bacon was truly devoted to the Christian faith.[142] He even wrote about his belief in a recent six-day creation: "Bacon further elevated Scripture above natural philosophy [science], considering the latter temporary and passing and the former divine and eternal."[143] Would a man like Bacon, who was truly devoted to Scripture – including a straightforward reading of Genesis 1-2 – have us eliminate God's Word entirely concerning origins? That does not seem reasonable. Bacon sought

---

https://creation.com/part-1-culture-wars-bacon-vs-ham

[141] Ibid.

[142] However, it has often been noted that Bacon may have been involved with both the Freemasons and the Rosicrucian's. This may or may not be true, but if true it may have been nothing more than a way for him to discuss the newest ideas in science and religion with like-minded scholars. In the time of the Enlightenment, these so-called 'secret societies' served a strong intellectual purpose. Sometimes they were attended by even the most devout Christian followers. They were outlets for those who wished to seriously ponder and debate meaningful issues of science, philosophy, and religion.

[143] Sibley, *Cracking the Darwin Code*, 47.

to limit God's Word in operational science only, but skeptics insisted that Bacon did not go far enough. The more radical skeptics of his time were convinced that we should neglect Scripture entirely, even in historical science. Some creationists have mistakenly believed that Bacon's scientific methodology "allowed no room for divine revelation,"[144] and within the area of operational science that is true. But this was not really a problem, as experimentation can be performed without any need of biblical input. As mentioned in the introduction, a Hindu, a Buddhist, a Muslim, a Christian, and an atheist can all perform experimental science without their worldview influencing the results in any way, and they can all arrive at the same conclusion. Scripture is necessary concerning historical science, however, since no experiments can be performed which shed indisputable light on how everything began. Yet as mentioned, skeptics 'seized the moment' and sought to remove Scripture from historical science as well. This helped paved the way for evolutionism's position of prominence among mainstream scientists in our time.

Andrew Sibley is a creationist who is extremely knowledgeable about the history of the origins debate. Fortunately, he sheds further light on this matter for us. It turns out that Bacon, who was countering a heresy in his time, was simply misunderstood:

> Bacon was not seeking to claim that Scripture can say nothing about science, but to counteract the inappropriate use of Scripture by the school of Paracelsus where proponents were seeking to base all [historical and operational] natural philosophy on a symbolic reading of God's revealed word.[145]

This would indicate that the followers of Paracelsus were attempting to integrate an allegorized reading of Scripture even into experimental science, the branch of science which has no need of biblical influence. Scientific experiments work fine apart from both Aristotle and Moses.

Therefore, as believers we should keep in mind the distinction between experimental science and the origins debate. These two aspects of science are, in some ways, worlds' apart from each other.

The Galileo Affair

Over three centuries before Christ, Aristotle taught that the earth was the center of the universe. This idea, known as 'geo-centrism' ('earth-centered'), persisted for centuries. In the second century AD Ptolemy further developed this idea, formulating what became known as the Ptolemaic system.

Then, in the sixteenth century the Polish astronomer Nicolaus Copernicus (1473-1543) advanced the idea that the earth and the other planets in our solar

---

[144] Sarfati & Wieland.
[145] Sibley, *Cracking the Darwin Code*, 44.

system revolved around the sun.[146] Since the ideas of Aristotle and Ptolemy held sway for centuries, this was quite a challenge to the academic and even religious establishments, for the Roman Catholic Church had accepted geo-centrism as well. Spiritually speaking, the idea that the earth is the center of the universe made sense to most Bible believers. In fact, Christians have long referred to verses that support the idea of geo-centrism (1 Samuel 2:8; 1 Chronicles 16:30; Psalm 93:1; 104:5; Ecclesiastes 1:5).

A century later, Galileo Galilei (1564–1642), relying upon observations afforded by his advanced telescopes, confirmed that Copernicus was correct. "For example, he observed that the sun had spots which moved across its surface, showing that the sun was not 'perfect' and it itself rotated; he observed the phases of Venus, showing that Venus must orbit the sun; and he discovered four moons that revolve around Jupiter, not the Earth, showing that the Earth was not the center of everything."[147]

GALILEO

'Helio-centrism' ('sun-centered'), better known as the Copernican system, was not the majority belief, however, as many scientists favored the teachings of Aristotle and Ptolemy regarding the structure and function of the solar system. These scholars were not always willing to change, and many of them either ignored or ridiculed Galileo's ideas. Many Church leaders were also persuaded that geo-centrism was taught in Scripture, and therefore Galileo was contradicting the Bible by teaching this position. To make matters worse, Galileo was often his own worst enemy. He was known to ridicule both scientists and clergy who opposed him.

Cardinal Robert Bellarmine was the most well-respected Roman Catholic theologian of the time, and he was also extremely knowledgeable concerning astronomy. This made him an especially important voice in the matter. In his opinion, the Bible probably did teach geo-centrism, but nonetheless he maintained that, if helio-centrism could be scientifically demonstrated, he would reconsider how Scripture should be interpreted in this matter. But Galileo had not sufficiently demonstrated the fact of geo-centrism:

[146] This was not a new idea, however. Aristarchus of Samos, a Greek astronomer and mathematician who lived in the third century BC, presented the first known heliocentric ('sun-centered') model shortly after Aristotle's time.
[147] Russell Grigg, "The Galileo 'Twist.'" https://creation.com/the-galileo-twist

> Galileo had not presented this [geo-centric] proof because he didn't have it. Instead, he bluffed, ridiculed his opponents and, in remarkable arrogance, claimed that the problem lay in their inability to follow his arguments. He became impatient and, early in 1616, sought to convince the pope (that is Pope Paul V), presenting the oceanic tides as 'proof' that the earth is not stationary. The pope responded by summoning his advisors, whom he asked to consider the matter. Under pressure to provide an answer, they responded quickly. Their conclusion was that Galileo's belief that the sun is stationary was contrary to Scripture and heretical, and that his view that the earth is not stationary was an error.[148]

If Galileo had humbled himself and cooperated with his opponents, things may have been drastically different. His arrogance made it easier for Church authorities to oppose him, forcing him to refrain from teaching helio-centrism unless it could be demonstrated beyond a shadow of a doubt.

Not surprisingly, skeptics are quick to pounce on their version of the 'Galileo Affair.' Their claim: Galileo, a humble scientist who followed the evidence where it led, was correct about geo-centrism, and the closed-minded, idea-stifling Church was flat-out wrong. So why not believe that secular scientists today are right about evolutionism while the Church is, once again, wrong? Even apart from the historical facts, there are some problems with this notion. First, there is nothing inherently atheistic about the earth either being or not being the physical center of the universe. (From a scriptural perspective, the earth does appear to be at the center of the universe, and it is also possible that our planet may be located at or near the physical center of the cosmos. But that is a different story.) On the other hand, evolutionism provides for an atheistic explanation of origins. Compared to the worldview implications of evolutionism, geo-centrism is not at all threatening to the belief in God and Christian salvation.

Second, science has long proven helio-centrism beyond a shadow of a doubt, and this concept is not at odds with Scripture. On the other hand, molecules-to-man evolutionism is both contrary to science and Scripture, leading people away from the one true God of the universe. Third, many Church leaders have accepted as fact non-Christian philosophies that are either blatantly anti-God or can easily lead to secular conclusions. Many Church leaders did this with geo-centrism in the days of Copernicus and Galileo by accepting as fact the ideas of Aristotle and Ptolemy without question. As we saw in chapter one, evolutionism is an ancient pagan idea that goes back even further than the Greeks. Yet many in the Church today accept evolutionism as fact. These theologians are using non-biblical ideas to interpret Scripture, rather than using Scripture to evaluate the perceived knowledge of the day. We will do well to avoid that mistake.

---

[148] Dominic Statham, "The Truth About the Galileo Affair." https://creation.com/galileo-church

The French Skeptics

During the Enlightenment, France was especially known for producing many atheists and hardened Bible skeptics. Although some of the skeptical were more deistic than atheistic, France saw a marked shift in pure naturalism (atheism) as the dominant worldview among the so-called 'scholarly elite.' This atheism was held in check only by the efforts of the Roman Catholic Church. Evolutionary thinking was 'all the rage' during this time in France. The prolific creationist author Jerry Bergman notes that, "The modern theory of biological evolution probably was first developed by Charles De Secondat Montesquieu (1689-1755)"[149] of France. As we have seen, evolutionism had already been around for a long time, not just centuries but millennia before Montesquieu, but in terms of a more modern system of evolutionary thought he preceded not only Charles Darwin but even Charles' grandfather Erasmus, who – as we will see shortly – was the 'real Darwin' behind Darwinism. Montesquieu maintained that in the beginning of life on the earth there were very few species, and they increased through biological change over eons of time. Charles Darwin expanded upon Montesquieu's ideas a century later.

Benoit de Maillet, a contemporary of Montesquieu, was another evolutionary thinker who greatly impacted the origins debate in his time. Among the French Enlightenment philosophers, he "was the first to propose a purely naturalistic view to the question: Where did we come from?"[150] Among other things, he taught that the universe was eternal, and that fish were the evolutionary forefathers of birds, mammals, and even human beings. (Remember that Anaximander taught this very same thing, six centuries before Christ). His writings were deemed so godless that even Voltaire, the deist who was quite possibly the most well-known skeptic of the French Enlightenment, opposed Maillet's teachings![151] That speaks volumes about how radical Maillet was in his time. Maillet was heavily influenced by Hinduism, which in large part explains his belief in both molecules-to-man evolutionism and an eternal universe. Of all the French evolutionists in the seventeenth and eighteenth centuries, he was the most influential – and the most radical.

Pierre Louis Maupertuis wrote in 1751 that new species may result when different anatomical parts recombine over vast eons of time, echoing the thoughts of Empedocles more than two millennia earlier. Around the same time, Denis Diderot taught that all animals evolved from one animal in the primordial past. Diderot began his academic career as a deist, but eventually slid into full-blown

---

[149] Bergman.

[150] Herbert, 75.

[151] Throughout the Enlightenment era, there was always a rift between the deists and the naturalists. The deists, although skeptical of the Bible's claims, could nonetheless see that a Master Designer was necessary to explain some things. For example, the distinct creation of human beings was often maintained by deists. Many of the deists were utterly repulsed by the thought of purely material causes for mankind.

atheism. "There was no God in Diderot's system of nature, only mindless accidents."[152] This was a totally Epicurean idea, for a new time and place. George Louis Buffon, a contemporary of Diderot, claimed that apes and man shared a common ancestor. Although this was an idea that Charles Darwin and others in his time would advance greatly, it is interesting that the foundation for human evolution was laid out a century earlier in France. (As we will see in chapter six, the idea of the 'ape-man' was ancient as well.)

The last major name in evolutionary thinking among the French was Jean Baptiste de Lamarck. His claim to fame was his theory of the inheritance of acquired characteristics. As an illustration of this idea, in the distant past a short-necked cow constantly stretched her neck to reach more edible leaves higher up on trees. As a result of constant stretching over many years, the cow's offspring were born with a longer neck. The original cow's neck lengthened only very slightly over time due to constant stretching, but this new trait was somehow passed on to succeeding generations, and this occurred for so long that the cows became an entirely new animal – namely the giraffe. Lamarck may have quibbled over the details of this explanation of his theory, but this is essentially what he taught. His theory was not even accepted in his time, let alone today.[153] (Actually, Charles Darwin was more accepting of Lamarckian ideas than was his grandfather, Erasmus Darwin. This shows us that Erasmus may very well have been the more reason-based scholar of the two, although they both 'missed the boat' considerably concerning origins.) Lamarck was a staunch atheist, so much so that his name became associated with atheism during his time – which says an awful lot, as atheism was prevalent among the French scholars of that day. Although biblical truth was being replaced more and more by the pagan evolutionism of the day, biblical creationism was still the dominant view of origins for the masses.[154] Therefore, Lamarck was not well-respected among the people, and sadly he died in obscurity.

Although the French Enlightenment scholars contributed greatly to evolutionary thinking in modern times, they are mostly forgotten today by the general public.[155] Many of the people discussed in this section are known only by those familiar with philosophy, history, and possibly the sciences as well.[156] Even today atheism remains a respectable worldview in France. This seems to be the sad legacy of the French Enlightenment scholars.

---

[152] Andrew Sibley, "Deep Time in 18th-Century France – Part I: A Developing Belief." https://creation.com/deep-time-in-18th-century-france-part-1/

[153] No one in their right mind today would make the claim that Lamarck was correct regarding his theory, since biology has advanced to the point that his idea has been thoroughly refuted.

[154] This was due to the still powerful influence of the Roman Catholic Church at that time.

[155] That is, unless you are French. The names discussed in this section still have recognition today in that nation.

[156] Although Voltaire's name is still somewhat recognized among the public today, even outside of France.

Erasmus Darwin: The 'Real Darwin' Behind Darwinism

In this 'Age of Reason,' perhaps the major name in evolutionary thinking was Erasmus Darwin (1731-1802), the grandfather of Charles Darwin. He was known for his great expertise as a physician, scientist, inventor, and philosopher. Although he authored several books, he expressed his ideas on evolutionism most notably in *Zoonomia* (1796). Erasmus was the 'true Darwin' behind Darwinism. It was Charles, however, who successfully packaged the theory and sold it to the world.

ERASMUS DARWIN

Erasmus was a renegade in his time, standing against the accepted institutions of Christianity and slavery as well as standing for the American and French Revolutions.[157] A poet as well, he penned his opinions and scientific thoughts in verse form in *The Botanic Garden*, *The Economy of Vegetation*, and *The Temple of Nature*. Among other things, he maintained that the earth was formed from a cosmic explosion, and that life began in the sea. (Once again, echoing the thoughts of Anaximander and other Greek philosophers' centuries before Christ.)

Zoonomia, his *magnum opus*, was so popular that it was translated into German, French, and Italian. Charles later admitted that he got much of the material for his theory from his grandfather, in large part through *Zoonomia*. *Zoonomia* truly was the original *Origin of Species*: "It has been called 'the first consistent all-embracing hypothesis of evolution,' and was published some 65 years before Charles published his version of evolution in *On the Origin of Species* in 1859."[158] In fact, "almost every topic discussed, and examples given, in *Zoonomia* reappears in Charles's major work [*Origin of Species*]. In fact, all but one of Charles's books have their counterpart in a chapter of *Zoonomia* or an essay-note to one of Erasmus's poems."[159] Yet few people today seem to be aware of Erasmus Darwin, the 'real Darwin' behind Darwinism!

---

[157] Erasmus Darwin's opposition to slavery, and his approval of the American Revolution, should be acknowledged as a credit to his character, despite his other beliefs.

[158] Russell Grigg, "Darwinism: It Was All in the Family." https://creation.com/darwinism-it-was-all-in-the-family

[159] Ibid.

The Rise of the Gap Theory

There have been many attempts over the years to harmonize the Genesis account of creation with secular geology, namely its teaching of 'deep time.' The gap theory was one of those attempts to reconcile millions of years with a straightforward reading of the Genesis account of creation.

Although I have already discussed the gap theory in the introduction of this book, let me quickly review it once more. Gap theorists insist that between Genesis 1:1 and 1:2, or possibly even between Genesis 1:2 and 1:3, are all the long ages needed for cosmic and geologic evolutionism, as well as the development of the earliest life forms on the earth. Instead of rendering Genesis 1:2 as, "Now the earth was formless and empty," they confidently assert that the verse should read, "Now the earth became formless and empty." Some great cosmic catastrophe, usually associated with the fall of Satan and the ensuing 'Lucifer's Flood,' destroyed the primeval earth. After long ages had passed, in which the earth laid in ruins, God re-created everything as outlined in a straightforward reading of Genesis 1. Essentially, this view combines either theistic evolutionism or one of the old earth creationist views (pre-catastrophe earth) with recent creationism (re-created earth). Therefore, some adherents of the gap theory instead prefer the term 'ruin-reconstructionism' rather than the gap theory.

Thomas Chalmers (1780-1847), a notable Scottish theologian and the first moderator of the Free Church of Scotland, was perhaps the man most responsible for the gap theory. A primitive version of the gap theory can be traced back to the rather obscure writings of Episcopius (1583-1643) as well as William Buckland (1784-1856), a geologist who was also an ordained clergyman. But Chalmers is the man who really popularized the idea. The most notable nineteenth century writer to promote this view was G.H. Pember, in his book *Earth's Earliest Ages*. Numerous editions of this work were published, with the fifteenth edition appearing in 1942. In the twentieth century the gap theory was heavily promoted through the work of Arthur Custance in *Without Form and Void*. There is still an abundance of biblical scholars today who hold to this view.

There are two major problems with the gap theory. First, Western Bible commentaries written before the eighteenth century – that is to say, before the commonly-accepted belief in an old earth became popular – knew nothing of any gap between Genesis 1:1 and Genesis 1:2. The gap theory seems to be a product of the Enlightenment era and even later, when scholars became interested in blending secular theories with Scripture. Second, the gap theory places death, disease, and suffering before the fall of humanity. (We will discuss this in more detail in chapter six.) Paul states, "Therefore, just as sin entered the world through one man, and death through sin, and in this way death came to all people, because all sinned."[160] Recent creationists insist that this verse refutes human sin

---

[160] Romans 5:12, NIV.

and death before Adam. The Bible teaches that Adam was the first human (Genesis 2:7; 1 Corinthians 15:45), and because of his rebellion death, disease, and suffering entered the world. Recent creationists are adamant that before Adam sinned, there could not have been any death – human or animal – and there certainly could not have been a race of people before Adam. Genesis 1:29-30 teaches that the animals and humans were originally intended to be vegetarian, a fact which is confirmed in Genesis 9:3-4. This is consistent with God's description of the creation as 'very good.' How could a fossil record which gives evidence of death, disease, and violence – after all, fossils have been found of animals apparently eating each other – be described as 'very good'? Therefore, we must assume that the death of billions of animals and even many humans, as seen in the fossil record, must have occurred after Adam's sin. The Noahic Flood provides an explanation for the presence of huge numbers of dead animals buried in rock layers which were laid down by water, all over the earth's surface. Paul states, "We know that the whole creation has been groaning as in the pains of childbirth right up to the present time."[161] Clearly the whole of creation was, and is, subject to decay and corruption because of sin, and the fossil record bears this out. Gap theorists believe that these horrors existed before Adam sinned, which seems to go against scriptural teaching.

Scientism: A Philosophy Long in the Making

Scientism refers to an excessive belief in the power of scientific knowledge and techniques. In one sense of the term, scientism is the belief that everything – or at least everything of importance – can be explained by science. This belief relegates philosophy and (especially) religion to the backseat, if not diminishing them altogether. The Christian apologist Dan Story comments on the increasing power of scientism in modern society:

> God and naturalism are mutually exclusive; if a creator God exists, naturalism as the foundation of secular science can't be true because it denies anything supernatural. So, it's in the interest (the very survival) of naturalism to deny God's existence and other religious truth-claims. The primary way it attempts to do this is by promoting science as the only reliable source of truth. In other words, if there is no transcendent, supernatural Being that reveals spiritual truth, it leaves only secular science to explain reality as it is. This philosophy is called 'Scientism,' and refuting it is our apologetic response.
>
> According to Scientism, nothing can be considered true – factual – unless it passes through the filter of scientific testing. If something can't be explained by modern science – that is, can't be proven scientifically – it can't be considered true or even

---

[161] Romans 8:22, NIV.

rational. Naturalists usually say something like this: "You can't know for sure if something is true unless you can prove it scientifically."[162]

Scientism and evolutionism have a lot in common. Science is good (or at least neutral, but not bad), while the philosophy of scientism is bad since it attributes too much explanatory power to a naturalistic-only interpretation of science. Likewise, evolution is good (or at least neutral, but not bad), while the philosophy of evolutionism is bad since it claims unproven and unsound beliefs as being beyond question. (To be fair, creationism should be questioned as well. This is not a problem, however, as creationists hope that people will question their worldview enough to consider the evidence for or against it. We welcome people to examine our beliefs.) Although scientism and evolutionism are not completely synonymous, the two beliefs are commonly held in unison by skeptics of the Bible.

The development of scientism took a long time. Millennia, in fact. This chapter will be a good place to discuss the history behind scientism, since much of it took place up to this point in our study. Plus, I want the reader to see how the foundations of scientism were laid out well before Darwin, which is where the following chapter begins. Many people have this mistaken notion that evolutionism, scientism, and a whole host of other 'ism's' began with Charles Darwin in the middle of the nineteenth century, but that is not even close to being true as we have seen already. Darwin is near the end of a long line of thinkers who paved the way for his ideas to flourish. He took a naturalistic system of origins, that had already been in place for millennia, and added some fresh ideas to it. (One of those 'fresh ideas,' natural selection, is scientifically solid. However, Darwin used it in the context of naturalism as part of his system of evolutionary thinking, even though creationists had already conceived of natural selection before him.)

Allan Chapman, who teaches the history of science at Oxford University, believes that scientism began with Auguste Comte in the early nineteenth century: "Comte effectively founded what we now often call scientism – or the worship of science and the scientific method as an infallible guide to absolute truth in itself."[163] Although I do not doubt that scientism in its current form began with Comte, it is safe to say that the foundations of scientism go much further back than this French skeptic. And a lot happened after him as well. As noted in chapter one, Anaximander of ancient Greece appears to have been the first anti-creationist that we know about. Although he may not have labeled his worldview 'scientism,' he almost certainly clung to that idea. All evolutionists, regardless of their time or culture, teach some form of science over religion. (For some

---

[162] Dan Story, "How Does Secular Science Attempt to Disprove God's Existence? (Part Four)." http://danstory.net/blog/

[163] Chapman, 46.

ancient cultures, there might have been an overreliance upon what we now know to be pseudoscience, but it was still their best science at the time.) Of course, there could have been earlier thinkers who promoted evolutionary ideas before Anaximander. Based upon Psalm 14:1, we can comfortably surmise that naturalism existed well before Anaximander. (David wrote the Psalms around 1,000 BC, over 400 years before Anaximander.) I strongly suspect that atheists were around before David as well, and where there is atheism there will be some form of evolutionism and an exaggerated reliance upon science (scientism). It might be good science, it might be unproven theories of science, or it might even be pseudoscience, but there will be some form of science that is attempting to replace biblical truth. Scientism has been around for a long time.

The Greek trio of Democritus, Epicurus, and Lucretius took evolutionary naturalism to a new level in their day, more than three centuries before Christ. Their version of 'science' is still with us today. The Epicureans in the early Church kept the first apologists on their toes, fighting against the idea of a chance, random creation in the distant past. Really, the ancient philosophy of atomism was not all that different from the evolutionary scientism of today.

The re-discovery of pagan evolutionism in the Middle Ages reintroduced Greco-Roman thinking to the West, and along with it came the ideas of Democritus, Epicurus, and Lucretius all over again. (Not to mention the evolutionism of Empedocles, Aristotle, and others.) Although it is well known that Aquinas favored Aristotle's teachings, we sometimes forget that Augustine showed great favor to the ideas of Plato, who was not exactly a biblical creationist. Islamic scholars may have discovered these pagan works first, but the Medieval scholars of the Church were quick to integrate these ideas into Christian theology.

Aquinas was committed to the teachings of Scripture, but nonetheless he placed great faith in man's ability to reason. This had to do with how he viewed the effects of Adam's fall on the intellect. Previously, Augustine contended that the fall of Adam severely impaired man's ability to reason. In fact, he was convinced that all aspects of man's character were severely corrupted by the fall. Aquinas disagreed, however. For him, God had given mankind two sets of gifts: Natural gifts and supernatural gifts. (Augustine recognized no such division.) Aquinas maintained that the supernatural gifts were lost at the fall, but the natural gifts were unaffected. Reason was the most important of the natural gifts, and since it was unaffected by the fall Aquinas contended that mankind was still able to reason his way to at least some truths – most important of all, God's existence as based upon nature.

Rene Descartes (1596-1650) further elevated reason. Although he was a man of faith, his philosophical system of rationalism paved the way for the heightened role of science over Scripture. Although we need both science and Scripture, we should never make the mistake of neglecting one over the other. During the Enlightenment, however, there were many thinkers who were quick to relegate faith to the background: Reason was 'in,' and faith was 'out.'

One of those radical thinkers of the Enlightenment was the Scottish philosopher David Hume (1711-1776). He denied miracles, including creation. Georg Wilhelm Friedrich Hegel (1770-1831), a later Enlightenment philosopher from Germany, was also critical of miracles. Both Hume and Hegel were very influential in the world of academia. Around this same time, the French skeptics of the Enlightenment were successful at fostering an attitude of religious skepticism. Both Voltaire (1694-1778) in France and Thomas Paine (1737-1809) in America paved the way for the nineteenth century biblical assault that was known as 'Higher Criticism.' (Or 'German Higher Criticism,' as it is oftentimes known, noting the nation of origin for this trend.) This was essentially Darwinism applied to the Bible.

Now we arrive at the time of Auguste Comte (1798-1857). His three stages of positivism promoted scientism in a major way. Positivism was the name given to his philosophical system which was intended to replace divine revelation with man's 'science.' The first stage was the religious or theological stage. In this stage, man invented the gods and supernatural beings to explain origins as well as all things mysterious or unknown. The second stage was the metaphysical stage. In this stage, man tried to discover his origins through philosophical speculations. This was considered a step up from the first stage, but still not sufficient for the 'noble beast,' man. The third stage was the scientific stage. In this final stage, man discovers existential truths through scientific observation and experimentation. In essence, man at this advanced stage sheds the superstition of his religious past, and he therefore becomes 'enlightened.'

The problem with scientism is twofold. First, to understand our origins we cannot neglect God's revelation. Science helps in man's understanding of his origins, but we cannot forget the revelation of the One who was there. Therefore, the first stage of positivism – the religious or theological stage – is man's most certain way of understanding his past origins. It turns out that the first stage is not the most primitive stage in man's development, but rather his most sure way of understanding his origins. Second, past events are not repeatable. Past events are historical in nature, not scientific. Therefore, science cannot tell us much about the origins of the universe, the earth, life, and humanity. Origins is an historical matter, not a scientific matter, and we need the help of the Great Historian to tell us how it all happened. Science can fill in the details of creation – in fact, a lot of the details – but God's revelation gives us the overarching plan of how it all began. We need both science and Scripture, not just science (scientism).

Scientism, or positivism as Comte would have preferred it known, was eventually applied to the Bible and faith. (This is the 'Higher Criticism' noted above.) Liberal theologians were quick to make this happen. They may have been professing Christians, and they may have thought they were helping their fellow man understand God, nature, and mankind's place in the universe, but their theology was anything but orthodox in nature. By the time the liberal theologians were calling into question the Christ of faith – eagerly replacing him with the Jesus of history, who was just a man and nothing more – special creation was already on the ropes, close to going down for the ten count. As in the ancient world, evolutionism was 'in' once again and creation by the hand of God was being pushed to the backseat.

David Friedrich Strauss (1808-1874) and Bruno Bauer (1809-1882) were among the first to write books about the historical Jesus from an extremely liberal (even naturalistic) position. Strauss taught that the miracles described in the Bible were nothing more than myths developed by early Christians to support their supernatural view of Jesus. In similar fashion, Bauer concluded that early Christianity owed more to ancient Greek philosophy, especially Stoicism, than to Judaism. He contended that Jesus was nothing more than a second century conglomeration of Jewish, Greek, and Roman theology. *Life of Jesus*, as each man titled his work, was nothing more than Socianism in an updated dress. Socianism, named for the sixteenth-to-seventeenth century Italian humanists Lelio and Fausto Sozzini (Latin, *Socinus*), taught that Jesus was merely a man. A good man, perhaps, but just a man. As Unitarians, they denied the triune nature of God, the divinity of Jesus, and everything else miraculous about the Bible and the Christian faith. They were typical among the skeptics of the Enlightenment. Not surprisingly, Professor Ferdinand Christian Bauer taught both Strauss and Bruno Bauer. F.C. Bauer, in turn, was a student of Hegel. As a result of this unbroken chain of mentoring, all these men denied the miraculous. The moral of the story: Be careful who you choose as your instructor, as it can be difficult for man to 'think outside the box' much of the time.

Of course, it is a bit difficult to reconcile Christianity with the belief that miracles are not possible. Some people have claimed to do this, but they are left with a Christianity that is anything but Christian. The German Higher Critics demonstrated this to be true. Not to be outdone, Ernest Renan (1823-1892) decided that he, too, needed to write a book titled *Life of Jesus*. (This seemed to be a popular title among the liberal theologians of that time.) His conclusions were no different from the others, however, and their combined influence continued to force itself upon the world of Christian theology. (With Renan, we are beginning to cross over into the next chapter chronologically, but I will keep going with the history of scientism in this chapter, to show the reader that much happened not only before the time of Charles Darwin, but after Darwin as well.)

As if things were not bad enough in Christology, along came Rudolph Bultmann (1884-1976). Bultmann, who had been a professor of New Testament at the University of Marburg in Germany, was convinced that the Bible was nothing more than Hebrew mythology, and that the New Testament needed to be 'demythologized' to get at the truth behind Christianity. The problem is, if one 'demythologizes' the New Testament, what is left is a story that is not just watered down, it is of absolutely no value to anyone.

It goes without saying that if Jesus were nothing more than a man, let alone a fictional conglomeration of Jewish, Greek, and Roman theology, then he was not the Creator of the heavens and the earth. Only God could reconcile fallen man back to himself, and only God can defeat Satan. If Jesus is not the Creator-God of Genesis 1:1, then Christianity falls. What the liberal theologians did was attempt to refute the very worldview they were trying to make more palatable to unbelieving man, but unbelieving man was never interested in that version of Christ (1 Corinthians 2:14). One only needs the real Jesus: Every other version of the Lord is a waste of time.

There were many other German theologians during the eighteenth, nineteenth, and twentieth centuries that caused a lot of damage to Christology,[164] but the point is this: Their theology messed things up in a big way. They influenced not just the world of Christian scholarship, but even the average Christian who was not especially enamored with theology eventually felt the weight of their unorthodox teachings. It may have taken a while, but scholarship always finds its way into the pews given enough time. Liberal Christianity made big strides across the theological landscape, and we are still feeling its effects today in our culture.

The reason that liberal Christology developed in the first place was that, over time, science was 'in' and theology and faith were increasingly viewed as primitive superstition. Scientism was not just developing, but rather it was solidifying. As if things were not tough enough for biblical creationism, along came White and Draper's 'conflict thesis.' Andrew Dickson White (1832-1918) was an American historian who co-founded Cornell University, serving as its first president for nearly two decades. He had also served as state senator in New York. He was a man with a great reputation in society, but in terms of the relationship between science and faith he missed the proverbial boat by a mile. In his book *History of the Warfare of Science with Theology in Christendom* (1896), he claimed there had long been conflict between science and religion. That is simply not true, however. There may be conflict between biblical creationism and Epicureanism-Darwinism, but there has never been conflict between science properly understood and Scripture. (We will examine this topic in greater detail in chapter five.) His proposed conflict between science and religion existed in his mind, and the minds of the other skeptics in his time, but the conflict was not real. It never has been.

John William Draper (1811-1882) was an English-born and raised chemist, physician, historian, and photographer who lived his adult life in America. He is known for having produced the first clear photograph of a female face as well as the first detailed photograph of the moon, both photos being taken in 1839-1840. He was the first president of the American Chemical Society (1876-1877), and a

---

[164] It was not just German theologians who were exceedingly liberal during this time, however. There were English, American, and other European theologians who were quick to teach a naturalistic version of Jesus and the Bible during this time.

founder of the New York University School of Medicine. He was obviously a learned man who accomplished much, but not all his accomplishments were good. His book *History of the Conflict Between Religion and Science* (1874) promoted an intrinsic conflict between religion and science, even though that conflict is imaginary only.

At the same time as White and Draper, the English biologist Thomas Henry Huxley (1825-1895) – better known as 'Darwin's Bulldog' – did as much or more to promote scientism. 'Darwin's Bulldog on the Continent,' the German zoologist Ernst Haeckel, was doing the same. As you will see in the next chapter, Haeckel had no problem 'bending the truth' (massive understatement) to promote scientism. There were others as well, but all these men did their best to advance science over biblical truth, whether that involved constructing an imaginary 'conflict thesis' between science and religion, or outright fabrication of supposed evidence's in their favor.

To conclude this historical review of scientism, although it may not be incorrect to say that scientism began with Auguste Comte in the early nineteenth century, it is more accurate to say that the road leading to scientism was long. Like many 'doctrines of demons,' scientism perverts that which is good – science – and attempts to turn it into something with an anti-God agenda.

Scientism, however, cannot answer several key points within science and philosophy. First, the beginning of the universe generates many questions. Why is there something rather than nothing? How can we account for a universe with a beginning, without positing a Beginner? How can science even begin to address a one-time event, namely the so-called 'Big Bang,' which is historical in nature (not scientific)? Two, why are the laws of nature so orderly and predictable if the universe is supposedly the result of a chance, random event ('Big Bang') and chance, random processes in nature (evolutionism)? Three, how can we account for the obvious fine-tuning of the universe, as well as life on the earth, in a supposedly impersonal and undirected universe? Four, how can we account for the origin of life, when the Law of Biogenesis clearly states that life comes only from life? This seems to point us in the direction of a personal originator of life. Fifth, how can we account for the origin of consciousness? How could creatures living in a supposedly impersonal universe seem to be so incredibly personal in nature? Finally, how do we account for the existence of moral laws, which are clearly inherent in the fabric of humanity? If the universe is a cold, impersonal, undirected place with no one in charge, why is there morality, love, and a host of other immaterial-yet-personal characteristics? For further study on the topic of scientism, I highly recommend Steve Cable's "Atheist Myths and Scientism," which may be found at Probe Ministries (probe.org).

'There is Nothing New Under the Sun'

As we have seen, evolutionism was not much of a threat during the Middle Ages, as the Church of that time kept this ancient philosophy in check. However, that all changed when the ancient pagan philosophers, along with their evolutionary ideas, had been embraced at this time by Muslim and Christian scholars alike. During the Renaissance and (especially) the Enlightenment eras evolutionism began its rise to prominence. The French atheists, Erasmus Darwin, and others were promoting evolutionary ideas that were indistinguishable from those espoused by ancient pagan thinkers such as Epicurus. Evolutionism, the ancient 'doctrine of demons,' was alive and doing well once again. As Solomon wrote three thousand years ago, "there is nothing new under the sun."[165]

From the ancient world to the so-called 'Age of Science' in the 1800's, the world had been steadily prepared for Charles Darwin's theory of evolutionary development. Really, Darwin's ideas were nothing but a rehash of earlier philosophies that were not just centuries, but millennia old. But the world in Darwin's time was more than ready for his theory: Fallen man needed a system of thinking that would allow him to 'throw God and the Church aside,' so that he could live the way he had always wanted to live. It allowed people to justify many things, as we will soon see in the next chapter.

---

[165] Ecclesiastes 1:9, NIV.

CHAPTER FOUR

Evolutionism from Darwin to the Present

In this chapter we will see how the acceptance of evolutionism increased dramatically with Charles Darwin's classic work *Origin of Species*. As we have seen, the list of evolutionists preceding Darwin was long indeed. Yet, those evolutionists tended to be the 'scholarly elite' of their time. Usually, the masses accepted creationism in some form or another. But after Darwin evolutionism became the norm for most people, even for many within the Church. It is safe to say that evolutionism – whether atheistic, theistic, or pantheistic – has replaced the doctrine of biblical creationism in our culture. However, evolutionism as a scientific theory is wrought with problems, as we will see in the next chapter.

Evolutionism in the Nineteenth Century

Before discussing Charles Darwin, it is first worth mentioning his greatest influence, Charles Lyell (1797-1875). Lyell was a uniformitarian who found it necessary to destroy any vestige of the biblical Flood in geology.

CHARLES LYELL

Although uniformitarianism in geology, when taken to its logical conclusion, leads to strict evolutionism in biology, Lyell nonetheless was uncomfortable with Darwin's assertion that mankind had descended from the 'beasts of the earth.' For Lyell, evolution might explain the rise, and especially diversification of the animal kingdom, but mankind had a divine origin. He may have been a committed deist, but he still maintained the biblical teaching of mankind being created in God's image (Genesis 1:27). In fact, Lyell was very troubled by Darwin's eventual push toward a strict naturalism, as the evidence for a Master Designer who 'set it all in motion' was obvious to him. Even among skeptics, there are some who are thought to be a little more radical than others. (As we saw in the previous chapter, the ideas of Benoit de Maillet were too much for Voltaire to handle.)

Charles Darwin's *Origin of Species*

Charles Darwin published *On the Origin of Species by Means of Natural*

Selection, or the Preservation of Favored Races in the Struggle for Life in 1859. As we saw in the previous chapter, even apart from his grandfather he was not even close to being the first to develop evolutionary ideas in relatively modern times. The French skeptics (Montesquieu, Maillet, Maupertuis, Diderot, Buffon, Lamarck), Thomas Malthus in England, and Benjamin Franklin in America had proposed evolutionary ideas in the same era as Charles grandfather, Erasmus. Alfred Russel Wallace, a self-taught naturalist and spiritualist leader in England, published papers on evolutionary theory just prior to Charles, as did the English naturalist Edward Blyth.

Among Charles contemporaries, Blyth was perhaps the most accomplished scholar of the day. His 1835 and 1837 papers on natural selection influenced Charles greatly. Blyth was the first to develop the idea of natural selection. The creationist author Russell Grigg discusses this powerful influence on Darwin:

> Edward Blyth (1810-1873) was the man whose ideas probably influenced Darwin most. An English chemist and zoologist, Blyth wrote three major articles on natural selection that were published in The Magazine of Natural History from 1835 to 1837. Charles was well aware of these. Not only was this one of the leading zoological journals of that time, in which his friends Henslow, Jenyns and Lyell had all published articles, but also it seems that the University of Cambridge, England, has Darwin's own copies of the issues containing the Blyth articles, with Charles's handwritten notes in the margins![166]

Additionally, the American naturalist Asa Gray corresponded with Charles before and after the publication of *Origin of Species*, no doubt shaping Charles' ideas on natural selection. When the first Neandertal skull was discovered in Germany's Neander Valley near Dusseldorf in 1856 there were many who had proposed evolutionary explanations for the strange shape of the skull, despite *Origin of Species* not being published until three years later. Evolution had already been 'in the air' for quite some time.

According to Darwinian evolution, natural selection operating over eons of time is the key to 'upward' (molecules-to-man) evolution. However, it is a well-known fact of biology that natural selection only serves to keep a species strong, rather than causing one type of plant or animal to turn into another, completely different life form. The problem for Darwinism is that microevolution, which is small changes within a species, does not lead to macroevolution – that is, one kind of plant or animal becoming a completely different kind of plant or animal. In 1980 about 150 of the world's leading evolutionists gathered at the University of Chicago, for the purpose of considering whether microevolution can lead to

---

[166] Russell Grigg, "Darwin's Illegitimate Brainchild." https://creation.com/charles-darwins-illegitimate-brainchild

macroevolution. Their answer was a resounding 'no.' Darwinian evolution rises or falls on this premise, and many evolutionists today are questioning their theory simply because of microevolution's failure to lead to large-scale change.

CHARLES DARWIN

Regarding Darwin's worldview, it has often been debated whether he was a deist or an atheist. As a younger man, he seems to have been a Christian. Michael Flannery, an old earth creationist, notes that, "Darwin started as a devoted creationist, even quoting Scripture to admonish the salty seamen with whom he resided on HMS Beagle during the five-year voyage that would make him famous."[167] Although opinions vary among those who have studied the issue in detail, it seems likely that Darwin was a professing Christian as a young man, then shifted to deism over time, and later still slid into full-blown atheism. Some of this may have been attributed to the death of his daughter Annie, who passed shortly before her tenth birthday. The problem of suffering and death has been a formidable one, challenging the hearts and minds of even the most devout believers at times. Therefore, we must strive to become apologetically minded and unafraid to wrestle with the big questions of faith. (We will tackle this objection to the faith in chapter seven).

Although there are many areas of Darwin's life and thinking that we could explore, what is most relevant for our study concerns the metaphysical influences on his theory. Although evolutionists would like the world to believe that Darwin arrived at his scientific conclusions based upon nothing more than testable science and observation, this is never the case for anyone. No scientist is a 'blank slate' who is free from outside influences – including creationists. The question is, "Which worldview has the support of scientific laws, well-documented history, logic, and – at least for the faithful – Scripture?"

Besides his grandfather Erasmus, who was a thoroughgoing materialist if ever there was one, the younger Darwin was also influenced through his involvement in the Plinian Society, an "Edinburgh student group devoted to discussing a wide range of philosophical and scientific topics (some quite radical)."[168] Flannery notes the impact that this group had upon Darwin:

---

[167] Michael Flannery, "What Did Darwin Know and When Did He Know It?"
https://reasons.org/explore/blogs/todays-new-reason-to-believe/read/tnrtb/2009/08/14/what-did-darwin-know-and-when-did-he-know-it

[168] Ibid.

Darwin was exposed to philosophical materialism of the most radical stripe as a teenager in Edinburgh through his involvement with the Plinian Society. Many of the topics he heard as a society member – man and beast as one, mind as matter, renunciation of natural theology, scientific positivism, philosophical materialism – turn up as themes in his later works. Based upon these facts, it seems likely that Darwin, perhaps initially unwittingly, filtered his observations through this Plinian worldview. Support for this paradigm shift is bolstered by his notebooks, which give frequent evidence of materialism along with substantial interest in skeptical – even atheistic – writers such as David Hume and Auguste Comte. Darwin's efforts to downplay these influences in his autobiography now make obvious sense: If the above is true his theory of evolution would be merely a materialistic metaphysic driven by a series of biological speculations, hardly the exacting science he would have the public believe. In short, Darwin came to his 'science' by way of his metaphysic and not the other way around.[169]

That last sentence says it all: "Darwin came to his 'science' by way of his metaphysic and not the other way around." But that is not how science is supposed to work. Aristotle and the other philosophers of ancient Greece are criticized today for having reasoned their way to scientific conclusions, rather than having arrived at scientific conclusions through testable experimentation. Yet that is no different than making one's scientific theories conform to their metaphysical prior assumptions. "Creationists are guilty of that, too!" scream materialists. Yes, creationists do base their view of origins upon Scripture, which is a metaphysical worldview. Creationists fully admit that, or at least we should. But, as already mentioned, one's worldview needs to be supported by scientific laws, well-documented history, and logic to be considered viable. Creationists have that support, whereas materialists do not. (The next chapter will point this out quite well.)

Was Darwinism founded purely upon testable experimentation and observation, free from prior assumptions? Hardly. Darwin was filtering his observations of nature through a lifetime of influences, including his grandfather Erasmus Darwin and Robert Grant, an early mentor of sorts who was a Lamarckian materialist. There were other influences as well:

Darwin had been proposed for the Society by a militant materialist and recent Edinburgh graduate, William A.F. Browne…his paper on mind as little more than brain matter was considered so inflammatory that it was struck from the Plinian

---

[169] Ibid.

minute book…Another Plinian, William Greg, just as iconoclastic as Browne, insisted that there was no difference between animals and humans. He invited his audience to conclude that "the lower animals possess every faculty & propensity of the human mind."[170]

As a younger man, Darwin absorbed these materialistic ideas like bare wood absorbs paint for the first time. Darwin eventually came to accept those same naturalistic conclusions, or at least strongly considered them as being true. Our influences shape the course of our thinking more than we realize. For an evolutionist to become a creationist, as I did many years ago, is truly an act of God. Likewise, for a creationist to become an evolutionist requires powerful influences opposed to God's clear witness in both Scripture and nature.

Let the Debates Begin!

The 1860 evolutionism versus creationism debate at Oxford College became the first of many formal debates between Darwinism and creationism. The English biologist Thomas Henry Huxley, known as 'Darwin's Bulldog,' challenged Samuel Wilberforce, the Bishop of Oxford. Not surprisingly, both men claimed victory! Although Wilberforce was a theologian, the initial resistance to Darwinism came from scientists rather than from theologians and pastors, who oftentimes blindly accepted Darwinian ideas. We would not expect that to have been the case, and certainly many scientists of the time were surprised by the theologians and Church leaders who so quickly bought into evolutionary thinking. Today, we see both scientists and theologians accepting Darwinism, often without question.

With the onset of the 'Internet Age,' online debates between creationists and evolutionists – not to mention believers and skeptics in general – have flooded the internet. It is now possible, and in fact quite common, for relatively unknown apologists to debate even-lesser-known skeptics on the theism-atheism debate as well as any number of related topics. Some of these internet debates are very well done, featuring apologists who are undoubtedly up-and-coming defenders of the faith. Likewise, some of the skeptics are quite sharp, and represent the skeptical community well. It will be interesting to see in the future how many of these relatively unknown debaters go on to achieve some level of notoriety in their respective camps.

Noah's Flood in the Age of Skepticism

The Noahic Flood also figured prominently into the origins debate during this time as well. Louis Agassiz, a contemporary of Darwin, became the first person to promote catastrophism. This is the belief that a series of geologic catastrophes occurred, each followed by divine recreations. The Noahic Flood was, according

---

[170] Ibid.

to Agassiz, the last of the great catastrophes. Biblical creationists today believe that many catastrophes have occurred throughout history, but only the Flood in the days of Noah was global in scope.

I have not spent much time discussing the Noahic Flood in this book. The Deluge is not the focus of our study – the issue of molecules-to-man evolutionism versus special creationism is where our time and energy has been, and will be, spent – but we should take a few minutes to think about this world-changing event. The Flood is, after all, a major part of the origins issue today.

If the Flood really happened, what evidence for it would we expect to find? Although books have been written on this subject, and a world-class museum has been built to demonstrate its historicity, I want to mention three lines of evidence in its favor: Scripture, geology, and anthropology. They all speak to an event that was global in extent.

THE DELUGE

Theistic evolutionists and old earth creationists typically hold to a regional flood, not a planet-wide event. (Having said that, however, at least one of my mentors – who is an old earth creationist – believes that the Flood was global in extent. It would be a mistake to maintain that every old earth believer subscribes to a regional flood.) Don Stoner represents the old earth position concerning Noah's Flood:

> The flood was indeed a river flood…The language of Genesis allows for a regional flood…The parts of modern Iraq which were occupied by the ancient Sumerians are extremely flat. The

floodplain, surrounding the Tigris and the Euphrates rivers, covers over 50,000 square miles which slope toward the gulf at less than one foot per mile…Drainage is extremely poor and flooding is quite common, even without large rainstorms during the summer river-level peak (when Noah's flood happened).[171]

At this point let us look at the language of Genesis, to see for ourselves if the Flood was just a regional disaster:

> For forty days the flood kept coming on the earth, and as the waters increased they lifted the ark high above the earth. The waters rose and increased greatly on the earth, and the ark floated on the surface of the water. They rose greatly on the earth, and all the high mountains under the entire heavens were covered. The waters rose and covered the mountains to a depth of more than fifteen cubits. Every living thing that moved on land perished – birds, livestock, wild animals, all the creatures that swarm over the earth, and all mankind. Everything on dry land that had the breath of life in its nostrils died. Every living thing on the face of the earth was wiped out; people and animals and the creatures that move along the ground and the birds were wiped from the earth. Only Noah was left, and those with him in the ark.[172]

As I read this passage of Scripture, I am struck by one thing in particular: This is not the description of a local or regional flood! Scripture states (1) the ark was lifted high above the earth, (2) the waters rose exceedingly high, increasing greatly on the earth, (3) the waters covered the mountains 'under the entire heavens,' which would indicate all the mountains on the earth, not just some of them, (4) every living thing that moved on the land and breathed air perished, and (5) only Noah and his family survived the Flood. As far as interpreting Scripture goes, there is no escaping the obvious conclusion that the Flood was global in extent.

Geologically, we would expect to find lots of fossils buried in sedimentary layers all over the earth. Buddy Davis, in the popular creationist song Billions of Dead Things, says it better than I can: "Billions of dead things, buried in rock layers, laid down by water, all over the earth." There are sedimentary rock layers, which were formed by lots of moving water, across the entire globe – and these layers are littered with the fossil remains of every type of creature imaginable. Sometimes these fossil remains are jumbled together in massive 'fossil graveyards.' That is what we should expect to find – and that is what we do find.

---

[171] Don Stoner, "The Historical Context for the Book of Genesis," Revision 2011-06-06, Part 3: Identifying Noah and the Great Flood. http://dstoner.net/Genesis_Context/Context.html#part3
[172] Genesis 7:17-23, NIV.

Anthropologically, we should expect to find several Flood legends from around the world. If the event really happened – and it clearly did, as it is plainly revealed in Scripture – then we should expect to have several cultural variants of the Flood story. It turns out there are at least 200 Flood legends worldwide, and some say there could be twice that many if you dig hard enough. But only 200 legends are more than enough to make the point: This event really happened, and people have passed stories about it down through the ages, in accordance with their culture and language. Perhaps the best book on this subject is Charles Martin's *Flood Legends: Global Clues of a Common Event.*

Did the Flood cause every geologic feature that we see on the face of the earth today? That is certainly not the case. It is also not responsible for every fossil ever produced, neither is it the source behind every Flood legend ever conceived. I strongly suspect that some of the legends were inspired by massive regional or local floods in the distant past. But Noah's Flood would be responsible for a lot of what we see in the earth today, and what we read about in legends from the past. It had to be, as this was an earth-changing event: The real estate was rearranged big time during the Flood! This was not a local or even a tranquil flood, as these types of disasters do not make it into the pages of Scripture. (The Flood is described in three chapters, while the creation event itself only warrants two. This catastrophe was anything but a footnote in the Bible.)

I recommend *A Flood of Evidence*, by Ken Ham and Bodie Hodge, for further study on this topic. I would also recommend reading some of the old earth creationist resources on the Flood as well.[173] Look at the differing views, and then think deeply about it. Which view makes the most sense to you? I accept that the Flood was a global catastrophe. Even some people who are not biblical creationists in the conventional sense of the term, such as Graham Hancock, accept the evidence for a global catastrophe in the distant past as well.

Creationists & Evolutionists Locked in Battle

Other early modern defenders of biblical creationism included Eleazar Lord (1788-1871) and his brother David Lord (1792-1880), and Philip Henry Gosse (1810-1888). Opponents of not only biblical creationism but God's existence included Ludwig Fuerbach (1804-1872), Karl Marx (1818-1883), and Friedrich Nietzsche (1844-1900). Interestingly, Nietzsche doubted many of the tenets of Darwinian evolutionism, but nonetheless remained steadfast in his devotion to atheism. "Nietzsche said Darwin was wrong in four fundamental aspects of his theory."[174] First, Nietzsche maintained that small changes could not produce new organs. He was ahead of his time in noting that microevolutionary (small-scale)

---

[173] The following old earth resources offer chapters devoted to Noah's Flood: *A Biblical Case for an Old Earth*, by David Snoke; *Navigating Genesis*, by Hugh Ross; *Origins: The Ancient Impact and Modern Implications of Genesis 1-11*, by Paul Copan & Douglas Jacoby.
[174] Russell Grigg, "Nietzsche: The Evolutionist Who was Anti-God and Anti-Darwin."
https://creation.com/nietzsche-anti-god-anti-darwin/

change does not lead to macroevolutionary (large-scale) change. Instead, natural selection only serves to keep a species strong, not to create new ones. (Creationists have been saying the same thing since Darwin's time.) Second, Nietzsche was adamant that the weak always outlast the strong. He wrote, "the weaker dominate the strong again and again – the reason being they are the great majority, and they are also cleverer…Darwin forgot the mind."[175] Third, sexual selection is not the norm. Nietzsche noted, "We almost always see males and females take advantage of any chance encounter, exhibiting no selectivity whatsoever."[176] For Nietzsche, sexual selection was not so much about selecting a mate as it was about the chance opportunity to continue the species. Lastly, the transitional forms that connect one type of creature to another are absent. Nietzsche was correct to note that these so-called 'missing link's' were still that: Missing! And, over a 150 years later, they are still missing. Despite an overwhelming commitment to materialism, even Nietzsche could see the fundamental errors in Darwinian evolutionism. If only evolutionists today could be so open to the evidence against their philosophy.

Eugenics: When Evolutionary Thinking Goes Horribly Wrong

Darwin's theory took off like a rocket in the middle of the nineteenth century, setting the stage for an onslaught of evolutionary and secular humanist teachings. One of those who was famous for spreading Darwinism was Ernst Heinrich Philipp August Haeckel (1834-1919), who was "known as 'Darwin's Bulldog on the Continent' and the 'Huxley of Germany'…notorious as the scientist who perpetrated fraud upon fraud to promote the theory of evolution."[177] Haeckel was a dedicated evolutionist, but not an atheist. For him, God was nature, and vice-versa. In other words, Haeckel was sold on the idea of pantheism. Haeckel saw God as an impersonal force who creates through a series of evolutionary events, with nature having a built-in 'self-ordering principle' that guides everything along. (As we saw in chapter one, Aristotle beat Haeckel to the punch on that idea, by over two millennia.) Never mind that science demonstrates no such 'self-ordering principle' in nature. Pantheism is a way for some to maintain that evolution has a guiding principle behind it, yet at the same time they can deny the biblical God. That approach seemed to work for Haeckel quite well.

---

[175] Ibid.

[176] Ibid.

[177] Russell Grigg, "Ernst Haeckel: Evangelist for Evolution and Apostle of Deceit."
https://creation.com/ernst-haeckel-evangelist-for-evolution-and-apostle-of-deceit
Thomas Henry Huxley, who was Darwin's number one 'apostle,' already had the nickname 'Darwin's Bulldog.' Therefore, Haeckel became known as 'Darwin's Bulldog on the Continent' since Huxley was from England and Haeckel was from Prussia (Germany).

ERNST HAECKEL

His enthusiasm for spreading Darwin's teachings was nearly unbridled. Sometimes that meant offering lines of evidence for evolution that were anything but factual. He invented the idea of 'monera,' a form of microbial life which supposedly filled the gap between non-living matter and the first signs of life. Not surprisingly, Thomas Henry Huxley went along with this idea. Huxley named a then-recently discovered life form, believed to fill the gap from non-life to life, after Haeckel (*Bathybius haeckelli*). It was later shown by a chemist to be "nothing more than amorphous gypsum, precipitated out of sea-water by alcohol!"[178] This did not stop Haeckel, however. He kept his original proposition unrevised for decades, even after the scientific refutation of monera was widely known among those in the scientific community.

There were other frauds perpetrated upon the scientific community as well. Haeckel imagined an ape-man that was a missing link in the supposed (make that imaginary) human evolution series, naming this fictitious species *Pithecanthropus alalus* ('speechless ape-man'). The great Rudolf Virchow, the father of modern pathology and a Bible-believing Christian, thoroughly refuted Haeckel's ape-man fraud. No such missing link has ever been found, of course.

However, it was Haeckel's fraudulent drawings of various embryos at different developmental stages that was the proverbial 'icing on the cake.' Haeckel did everything he could to promote the idea that animals go through various evolutionary stages in their embryonic development. For example, the human fetus must go through the 'fish stage,' then the 'amphibian stage,' then a 'lower mammalian stage' of some sort, and so on before resembling a human being in the womb. This reinforced the humanistic idea that man is really nothing special, being just another animal originating through a long series of chance, random accidents of nature. This fictitious series of embryonic development serves as an icon for molecules-to-man evolutionism even today. For some people, the idea of 'helping along' human evolution would be advantageous to our species. The weak are eliminated, leaving only the strong. But is that how God intends for us to treat our fellow man? Hardly.

Not surprisingly, Haeckel's fraudulent idea – known scientifically as 'ontogeny recapitulates phylogeny' – can easily lead to the idea that mankind

---

[178] Ibid.

must control his future by eliminating the so-called 'weak and unfit.' And who might that include? The physically and mentally disabled, as well as those belonging to various ethnicities or people groups of some sort. If one group of people decides that another group of people are 'less than desirable' in some way, then maybe that undesirable group should be eliminated. One can easily see where this idea is headed. As we already know full-well, this situation has happened countless times throughout mankind's history.

Unfortunately, Haeckel's evolutionary ideas partly led to the "intense German militarism"[179] that fostered World Wars I and II, as well as the Holocaust. Eugenics or 'controlled evolution' took the work of others as well, such as the German atheist Friedrich Nietzsche and even Charles Darwin's cousin, the eugenicist pioneer Francis Galton. The path to genocide was set in motion through the warped evolutionary thinking of these men, and soon after their time another man – one who can only be described as 'pure evil' – came along and put this idea of 'evolutionary cleansing' into practice.

Adolf Hitler admired the work of Charles Darwin, Ernst Haeckel, Friedrich Nietzsche, and Francis Galton. Like Haeckel, Hitler was a pantheistic evolutionist: "In his infamous *Mein Kampf*, Hitler uses 'creator' with reference to nature."[180] It is worth noting, however, that pantheism and atheism are synonymous in that pantheism denies the transcendent Creator who possesses the attributes of personality, as does atheism. Regardless of the distinction, however, Hitler was fully committed to evolutionism, supporting the ideas of distinct races evolving at different rates and 'survival of the fittest.' This type of thinking led him to believe that some races were superior to others – and that something had to be done about the 'inferior' races.

HITLER

Hitler believed that the ideal human being was of a Nordic-Germanic stock. Unfortunately, he saw almost everyone else on the planet as being inferior in some way – especially the Jews. Over six million people died in the Holocaust that was initiated by Hitler, and the vast majority of those murdered were of Jewish ancestry. In his warped mind, he was simply 'helping evolution along' by getting rid of the weak and undesirable. But who decided who the 'weak and undesirable' were? Hitler, of course.

Hitler failed to understand that all people are created in God's image, not just

---

[179] Ibid.
[180] John Woodmorappe, "Hitler the Evolutionist; Hitler the Pantheist (Hitler the Atheist – Yes)." https://creation.com/review-hitlers-religion-weikart

some of us. Sadly, we live in a world that has not always figured that out. (It often feels like our modern culture is totally ignorant of this fact.) To better understand the roots of racism, and how it fits into the history of the evolutionism-creationism debate, we need to explore the Great Chain of Being. In ancient Greece, the Great Chain was the accepted hierarchy of everything, whether spiritual, animate, or even inanimate. The concept began with Plato and his student Aristotle, but it saw further development during the Medieval era. The Great Chain began with God and the spirit realm, then extended to mankind and earthly things. The hierarchy was as follows: God, angels, mankind, animals, plants, and minerals. Additionally, except for God each entity consisted of subdivisions. Therefore, the Great Chain became quite complex. Angels were ranked in importance, according to their type or 'office,' as were the animals, plants, and even the inanimate things of this world (gems, rocks, soils, etcetera). Sadly, even mankind was ranked at various times in history. For example, during the Medieval era the king stood at the top of the Great Chain, followed by the aristocracy, the clergy, the skilled laborer, and last of all the peasant. Unfortunately, many people throughout history have also ranked human beings according to 'race' on the Great Chain. Not surprisingly, the white European always stood higher up on the Great Chain, while the other people groups were ranked somewhere lower down. However, Paul wrote that God created every nation of mankind from only one man (Acts 17:26). Therefore, this idea of ranking the so-called 'races' of man according to imagined superiority is an unscriptural – not to mention unscientific – idea. Like evolutionism, it is nothing more than a 'doctrine of demons' intended to create hate and division. It should come as no surprise that racism and evolutionism appear together so often in history.

Table 4.1

Great Chain of Being

GOD
↓

ANGELS (RANKED BY TYPE OR OFFICE)
↓

MANKIND (SOMETIMES RANKED BY SOCIAL STATUS OR RACE)
↓

ANIMALS (HIGHER ANIMALS-TO-LOWER CREATURES)
↓

PLANTS (GREAT TREES-TO-THORNS, THISTLES, WEEDS)
↓

MINERALS (GEMS, ROCKS, SOILS)

In a similar way, the idea of molecules-to-man evolutionism is represented by the drawing known as the 'Tree of Life' which appeared not long after Darwin wrote *Origin of Species*. In the Tree of Life drawing, a 'simple' cell lies at the bottom of the page. (There is no such thing as a 'simple' cell, but some cells are less complex than others in some ways.) Above this cell are several branches of living things, both plants and animals. The idea behind the drawing is this: Every living thing, be it plant or animal, evolved from that simple cell over eons of time.

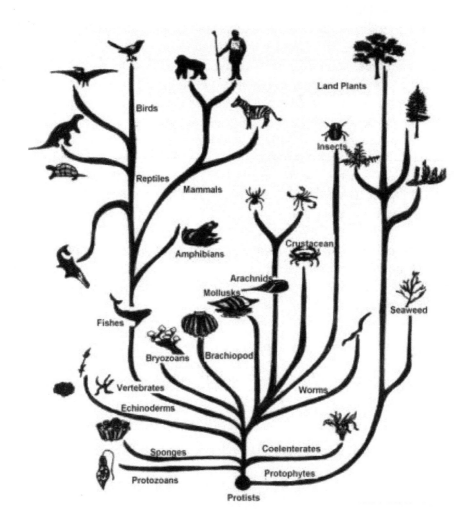

TREE OF LIFE

Concerning the animal kingdom, the 'lower' life forms are first, lower down on the page, while the 'higher' forms are found above that. Once again, the idea behind this artistic expression is that life evolved from simple to more complex forms over time. For example, archaic marine creatures preceded the land vertebrates, which eventually gave rise to the great apes. The problem with the

Tree of Life is that there is a myriad of missing links between the various life forms. Evolutionists are simply not sure how many creatures are supposedly linked in any given 'branch.' Of course, there are no missing links, which is why there are so many unknown transitions shown on this imaginary drawing. Instead of the Tree of Life, plants and animals give every indication of being created distinctly, with some variation within a created kind (microevolution). Therefore, instead of a Tree of Life, we really have an 'Orchard of Creation.'

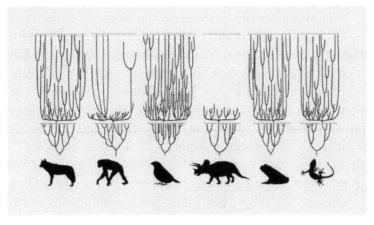

ORCHARD OF CREATION

The main point is this: The Tree of Life is like an inverted Great Chain. The Tree of Life begins at the bottom of the drawing and everything progresses upward from there, while the Great Chain begins with God at the top of the drawing, and everything regresses downward from there. Of course, the Tree of Life makes no mention of God, as the divine may or may not be part of the equation for evolutionists. Interestingly, many people have ranked man according to 'race' on both the Great Chain and the Tree of Life. Creationists, however, have always known that man is man, regardless of skin color or any other physical characteristic. Hitler may or may not have been familiar with, or even cared about, the Great Chain. If he did incorporate this idea into his belief system, his version of the divine – the impersonal god-force of pantheism – would have been at the top of the hierarchy, while mankind would have begun with the Nordic-Germanic people followed by a series of differing ethnicities. (The Jew would have occupied the lowest rung of the ladder in Hitler's twisted hierarchy.) It is certain, however, that Hitler was familiar with the Tree of Life, and surely at the top of the human heap for him was the white European. (Especially those of a Nordic-Germanic stock, over against the Italians, Spaniards, Slavonic peoples, and so forth.) Whether it is the Great Chain or the Tree of Life, racist thinking has found its way into both hierarchies. Fallen man always thinks that way.

The Rise of Theistic Evolutionism in the Church

Once *Origin of Species* became influential in the late nineteenth century, it was almost immediately that the theologians and leaders of the Church began to integrate evolutionism into the doctrine of divine creation. Initially the resistance to Darwinian evolutionism came from scientists, while theologians seemed content to make evolutionism part of Genesis creation.[181] James McCosh (1811-1894) may have been the most prominent of the early post-Darwin theistic evolutionists.[182] Through his influence upon the world of Christian theology as the President of Princeton College (1868-1888), theistic evolutionism was granted a certain degree of respect. McCosh was convinced that Christianity and evolutionism could be reconciled, with no loss of integrity to either science or Scripture. In his defense, he did not have access to the latest discoveries in microbiology, genetics, and cosmology that we do today, so his eagerness to accept evolutionary ideas is somewhat forgivable. McCosh became the first voice for Darwinian evolutionism in America by a religious leader.

Fortunately, McCosh was opposed by many theologians and Church leaders of the time, as not everyone was willing to accept evolutionism. Charles Hodge, perhaps the greatest theology professor contemporaneous with McCosh, considered Darwinism to be blatantly atheistic. Since *Origin of Species* was published in 1859 and Hodge passed in 1878, he was the first prominent theologian to oppose Darwin's ideas, although he only had the chance to do so for a few decades.

Hodge's biblical position on creationism held sway until the time of his death, but not long after that evolutionary teachings began to gain much ground in the Church. (Sometimes all it takes is one good leader to stem the tide of bad thinking. But when that leader is gone, be on guard for what happens next.) Hodge's creationist followers were supplanted by theistic evolutionists in 1929, and eventually Princeton College – now Princeton University – became known as a leading institution for evolutionary biology. Sadly, the number of colleges, universities, and even seminaries that teach special (non-evolutionary) creation is dwindling worldwide. Many seminaries today are teaching only theistic evolutionism.

Today, theistic evolutionism enjoys a position of prominence among Christian believers through the efforts of BioLogos, a Christian-based "think tank"[183] which promotes the melding of Darwinism and Scripture through an

---

[181] Of course, this is a general statement. There were some scientists who accepted Darwinism without question, while some theologians fought against evolutionary ideas. But in general, this was the surprising trend in the latter half of the nineteenth century.

[182] The great Princeton Seminary theologian B.B. Warfield (1851-1921) is often believed to have been a staunch proponent of theistic evolutionism, but his acceptance or denial of this belief is debatable.

[183] John Upchurch, "The Danger of BioLogos."
       https://answersingenesis.org/theistic-evolution/the-danger-of-biologos/

appeal to divine guidance of natural processes. BioLogos originated through the efforts of several people, but perhaps more than anyone else it was Dr. Francis Collins, an American physician-geneticist who headed the Human Genome Project. He directs the National Institutes of Health in Bethesda, Maryland and is the author of *The Language of God*, which combines his Christian testimony with his views on science and evolutionary principles. Theistic evolutionists such as Dr. Collins are truly Christian brothers and sisters whose commitment to Christ should not be in question. The problem is not personal in any way. He, along with the other scientists and theologians associated with BioLogos, are fine examples of professing believers who serve the Lord wholeheartedly. It is simply that theistic evolutionism has some serious faults. However, if there is one single argument against theistic evolutionism, it is this: It requires death before the fall of mankind. At the conclusion of the six-day creation week, God pronounced his creation as 'very good' (Genesis 1:31), since the fall of mankind had not yet taken place. It is troubling for many to believe that God would call his creation 'very good' if death, disease, and suffering had already entered the world by the time of Adam and Eve in the Garden of Eden. (According to theistic evolutionists, it took billions of years of evolutionary change to culminate in human beings. This includes a long chain of 'sub-humans,' despite Scripture clearly noting that Adam was the first man. Evolutionism really is a philosophy of death.) Paul wrote that, "sin came into the world through one man, and death through sin, and so death spread to all men because all sinned." Sin did not enter the world before Adam and Eve's rebellion,[184] yet theistic evolutionism requires death on an unimaginable scale before Adam to bring about the various life forms over vast expanses of time. Did Adam and Eve enjoy God's 'very good' creation in the Garden of Eden, all the while standing upon layers of dead and buried creatures? Theistic evolutionists must believe this. This is the fundamental point of contention between recent creationists and their old earth brethren, be they old earth creationists or theistic evolutionists.

Evolutionism in the Twentieth Century

It is believed by some people that the Canadian George McCready Price established the modern young earth creationist movement early in the twentieth century. However, recent creationism was really nothing new at the time. As we already saw, most Church Fathers maintained a belief in a literal six-day creation just thousands of years ago, as well as a global Flood in the days of Noah. It was the 1961 publication of *The Genesis Flood*, by American creationists Henry Morris and John Whitcomb, which became the springboard that launched worldwide interest in recent creationism. Although it had already been established many centuries earlier, the doctrine of biblical creationism had suffered serious suppression from the time of the Enlightenment onward. Several

---

[184] Death before Adam's fall is a huge area of debate between young earth and old earth creationists. We will look at this issue in more detail in chapter six.

creationist organizations formed shortly after this game-changing publication by Morris and Whitcomb, including the Creation Research Society in 1963, the Institute for Creation Research in 1970, the Creation Evidence Museum in Glen Rose, Texas in 1984, and Answers in Genesis' American headquarters in 1994, which originally began years earlier in Australia. Their old earth competitor, Hugh Ross, founded Reasons to Believe in 1986, while the theistic evolutionist group BioLogos was founded relatively late, in 2007. The debate over biblical creationism is not just confined to believers and skeptics, as it is an intramural debate as well. In this regard, we should all be like the Bereans, who studied the scriptures intently to determine the validity of certain beliefs (Acts 17:10-11).

The Scopes Trial in 1925, held in Dayton, Tennessee became the most famous court case of its kind. The Butler Act prohibited the teaching of evolutionary theory in Tennessee, yet John Scopes was found guilty of teaching evolutionism to his students and was fined for doing so. The case was later dismissed on a technicality, but it served its purpose of challenging the Butler Act.[185] Shortly after the Scopes Trial a former student of George McCready Price coined the term 'creationist.' Prior to this time, creationists had been referred to as both 'Christian fundamentalists' and 'anti-evolutionists.'

The American Humanist Association formed in 1933, in part as a response to the continuing debate on origins. The original members were avid promoters of an eternal, self-created universe and the naturalistic evolution of life. Although most evolutionists have given up on the belief in an eternal universe, they still hold tenaciously to the belief in the spontaneous generation of life from non-life and molecules-to-man evolutionism.[186] The Humanist Manifesto I, written in the same year that the organization was formed, served as the 'statement of faith' for all atheists worldwide. The Humanist Manifesto's II (1973) and III (2000) continued to encourage atheists throughout the years. As we have seen, the battle for origins began in the very distant past, far back in the ancient world, and it shows no signs of letting up anytime soon.

The Rise of Pantheistic 'New Age' Evolutionism

There is another version of evolutionism besides the atheistic and theistic varieties that we have discussed so far in this chapter. In the first half of the twentieth century, evolutionism gained prominence through the efforts of Teilhard de Chardin (1881-1955), a Roman Catholic priest who spent far more time engaged in the search for human origins than attending to priestly duties. He spent a great deal of time searching for mankind's supposed evolutionary ancestors, the so-called 'ape-men.' He spent years excavating in China, where he took part in the Peking Man excavations in 1929. Unfortunately for Chardin,

---

[185] The challenge was initiated by the American Civil Liberties Union (ACLU), which has had great success in bolstering the efforts of atheists in this country.

[186] In the naturalistic evolutionary worldview, spontaneous generation had to have occurred at least once in the very distant past.

he was also associated with the Piltdown Man hoax in 1912. When Piltdown man was proven fraudulent in 1953 – amazingly, it took scholars 41 years to figure out this was a fraud – Chardin was embarrassed by this stain on the science of paleoanthropology. Piltdown man showed the world that scientists are only human, and they make mistakes like everyone else. Maybe, just maybe, they do not have everything figured out after all.

TEILHARD DE CHARDIN

Unlike the evangelical theologian James McCosh in the previous century, Chardin saw no possibility of harmonizing Genesis with evolutionary ideas. Genesis 1 became nothing more than outdated Hebrew mythology for him, at best a primitive attempt to explain origins apart from the 'sophisticated' science of evolutionism. He "believed strongly that a Church which accepted the Genesis account of creation was wedded to an outmoded and unscientific outlook. In his framework, such a Church was out of touch with reality and would lag behind the rest of the world as it plunged into the twentieth century."[187] How many pastors and seminary professors today make that claim as well?

Many scholars and historians have debated whether Chardin was a pantheist or not. Based upon his writings, it seems likely that he was. Many pantheists believe that there is a universal reasoning ability or 'Universal Mind' that pervades the universe, while the physical universe itself can be thought of as being God's body. Many of the ancient Greek philosophers believed this, as do some of the Eastern religions. Another Eastern religious idea that Chardin held to is the notion that the created universe is not real. Instead, the only reality is that which is spiritual. (This is an idea that many of the ancient Greek philosophers, and the Gnostics of the early Church era, held to as well.) Chardin believed that the process of evolution was inherently intelligent:

> [Chardin] proposed that the universe did not consist of real things, since everything was evolving and converging towards a future goal called 'Omega.' The only thing that must give it unity, therefore, is the spiritual or mystical realm. God must be the only unifying force…God somehow inserted himself into the

---

[187] G.J. Keane. "The Ideas of Teilhard De Chardin." https://creation.com/teilhard-de-chardin

evolutionary process, and Christ the force drawing everything towards the goal of 'Omega.'[188]

Chardin did manage to work Christ into his belief system, but the Christ that he offered was nothing more than an evolutionary force that appealed to mystics like himself. The Roman Catholic Church was seriously challenged by Chardin's ideas!

Henry Morris remarked that, "Many of them [New Age adherents] regard the controversial priest, Teilhard de Chardin, as their spiritual father."[189] Morris also stated that the god of the New Age is 'Evolution,' capitalized to stress its great importance in New Age theology. Chardin was certainly not the only contributor to the modern New Age phenomenon, but his influence cannot be taken lightly.

Not just a biological evolutionist, Chardin also held to cosmic evolutionism. In *The Phenomenon of Man*, Chardin wrote the following:

> Some thousands of millions of years ago, not, it would appear, by a regular process of astral evolution, but as the result of some unbelievable accident (a brush with another star? an internal upheaval?) a fragment of matter composed of particularly stable atoms was detached from the surface of the sun. Without breaking the bonds attaching it to the rest, and just at the right distance from the mother-star to receive moderate radiation, this fragment began to condense, to roll itself up, to take shape. Containing within its globe and orbit the future of man, another heavenly body – a planet this time – had been born.[190]

Our planet is merely an accident? That is a long way from the, "In the beginning God created the heavens and the earth"[191] statement of faith that is supposed to be taught by Roman Catholic priests. When you get right down to it, Chardin's 'astral accident' theory is no different than the nebular hypothesis of Pierre Simon Laplace (1749-1827). Writing well over a century before Chardin, this fellow Frenchman taught that the primordial sun was a giant gas cloud or nebulae that rotated evenly. But as the gas contracted due to cooling and gravity, the cloud began to rotate faster. This increased speed of rotation threw off a rim of gas, and once it eventually cooled it condensed into a planet. Laplace claimed that this process was repeated several times, to produce all the planets in our solar system. Not every rim of gas that was ejected condensed, however. Some rims of gas became asteroids, such as those between Mars and Jupiter. The now smaller, but still massive, gas ball which remained in the center became our

---

[188] Ibid.
[189] Henry M. Morris, "Evolution and the New Age." https://icr.org/article/evolution-new-age/
[190] Teilhard de Chardin, *The Phenomenon of Man* (London, England: William Collins Sons & Co. Ltd., 1980), 73.
[191] Genesis 1:1, NIV.

sun. An interesting idea, but why did the primordial sun stop ejecting rims of gas after a certain point? It seems likely that it would have continued to do so until its' mass was completely gone, or at least until it was considerably smaller than it is now. This seems a little too convenient. While Laplace's "theory is based largely on the observation that all then known planets revolved around the sun in the one direction,"[192] which is true, not all the planets rotate on their axis in the same direction. Venus and Uranus rotate in a different direction from the others. Likewise, not all the moons in our solar system rotate their planets in the same direction. However, according to the laws of physics everything should be orbiting and spinning in the same direction – exactly as the original gas cloud supposedly did. The fact that there are differing orbital and rotational directions makes no sense in an evolutionary worldview. It is, however, no problem according to the creationist account of origins. Helmut Welke comments further on this phenomenon:

> Today, evolutionists insist the planets were formed from swirling gasses that also formed the sun. The physics of all these fast-moving gases would result in a sun that rotates faster than it does. The planets would all then orbit the sun in the same direction, and they do. But all the planets should rotate or spin in the same angular direction as they move in their orbit. Venus and Uranus spin the wrong way. (Venus in reverse, and Uranus spins horizontally to the main plane of the solar system.) This cannot be if they formed from the same cloud of swirling gasses. It is one of several problems with evolutionary thinking on the formation of our solar system.
>
> Of course, evolutionists have an answer for this problem. They propose an object or two collided with these planets in their early formation. This changed their rotation so much – to what we see today. But many people find this solution of planetary billiards to be extremely implausible. Perhaps the best explanation is that God chose to spin these planets very differently, just to confound anti-God evolutionists of today.[193]

Not every bad idea dies right away, however: "It is interesting to note that, apart from the Biblical account of creation this theory on the origin of the solar system has been adhered to longer than any other. It still has a wide acceptance, and it is currently being promoted in modified form by theorists."[194] To the unregenerated man, anything but biblical creationism seems to ring true.

---

[192] John Mackay & T. Parsons, "Pierre Simon Laplace: The Nebular Hypothesis." https://answersingenesis.org/astronomy/cosmology/pierre-simon-laplace-the-nebular-hypothesis/

[193] Email exchange between Helmut Welke and myself, December 1, 2020.

[194] Mackay & Parsons.

When you get right down to it, evolutionism was Chardin's religion, not Roman Catholicism. Chardin wrote, "Is evolution a theory, a system or a hypothesis? It is much more: It is a general condition to which all theories, all hypotheses, all systems must bow and which they must satisfy hence forward if they are to be thinkable and true. Evolution is a light illuminating all facts, a curve that all lines must follow."[195] But as the evidence against evolutionism becomes stronger, with advances in microbiology, genetics, and cosmology, Chardin's view becomes relegated to the scrapheap of bad ideas. "Teilhard's speculative theories were not scientific, but metaphysical! They depended for plausibility upon evolution being historically true. As the credibility of evolution theory diminishes, his writings reduce to highly imaginative anti-Christian fantasy."[196] That is true not just for Chardin's version of evolutionism, but for the evolutionary worldview in general.

In his time, Chardin was considered a 'progressive' by secular-minded academics. He was a hero to many evolutionists in his time:

> [He became] a cult figure to many after his death, particularly to academic evolutionists among Catholics and Anglicans. Many still believe his ideas were ahead of his time, and that his thinking will inevitably be accepted by the official teaching bodies of the Catholic Church. The reality is however that his confused speculation has only contributed to further obscuring the notion that God has revealed objective truth to man through the Holy Scriptures. Mysticism has always resulted in common sense being replaced by nonsense.[197]

Sadly, way too many churches today have entertained pantheistic evolutionism such as that promoted by Chardin. Incorporating the imaginary 'god-force' into evolutionism seems to make this version more palatable for many evolutionary theorists today.

Chardin was not the only evolutionist of note from the pantheistic camp, however. Morris notes that both Julian Huxley (1887-1975) and Theodosius Dobzhansky (1900-1975), two of the most prominent neo-Darwinians in the middle of the twentieth century, were early proponents of an evolution-based mystical religion. Morris found this very surprising:

> In a eulogy following Dobzhansky's death, geneticist Francisco Ayala said: "Dobzhansky was a religious man, although he apparently rejected fundamental beliefs of traditional religion, such as the existence of a personal God...Dobzhansky

---

[195] Chardin, 241.
[196] Keane.
[197] Ibid.

held that in man, biological evolution has transcended itself into the realm of self-awareness and culture. He believed that mankind would eventually evolve into higher levels of harmony and creativity. He was a metaphysical optimist.[198]

Dobzhansky himself wrote, "In giving rise to man, the evolutionary process has, apparently for the first and only time in the history of the Cosmos, become conscious of itself."[199] Evolution became conscious of itself? If that is not a religious idea, then nothing is. Yet this very notion is so prominent today that it boggles the mind.

Harvard University biologist George Wald, a Nobel laureate, wrote, "The universe wants to be known. Did the universe come about to play its role to empty benches?"[200] The belief that 'the universe wants to be known' is an interesting idea, but it does not square with the thinking of countless evolutionists throughout history. If the universe is merely one big cosmic accident, as so many evolutionists insist, then there is no 'universal mind' that wants or expects anything. The universe is allegedly nothing more than mindless matter, so why should anyone make it out to be conscious of itself? I believe the reason behind this thinking is obvious: Pure naturalism (atheism) is an unsatisfying worldview. People inherently know that there is a spiritual realm beyond the physical cosmos (Ecclesiastes 3:11; Romans 1:20; 2:14-15), so the 'spiritualizing' of the creation is merely a way to recognize this spiritual realm without giving the one true God of the universe the praise and worship that he alone deserves.

The list of pantheistic evolutionists far exceeds those mentioned above. The idea that the universe is alive and capable of conscious thought is pure mysticism to the core. Yet that seems to be where we are today. A huge contributing factor in this trend is Hollywood, which has produced several movies promoting the belief in a pantheistic universe filled with mystical ideas. The *Star Wars* saga serves as the best example. These movies introduced us to 'The Force,' which is the consciousness of the universe that can be tapped into by the spiritually adept. Many movie stars and celebrities of all stripes have embraced mysticism in one form or another, and they are intent on bringing their ideas to the world. This pantheistic-mystical outlook on life gives the veneer of being spiritual or religious in nature, but ultimately this worldview points' people away from the truths found in Scripture.

As we have seen, evolutionism has branched out into three camps: Atheistic, theistic, and pantheistic. Not surprisingly, many apologists recognize that there are three dominant worldviews today: Atheism, theism, and pantheism.

---

[198] Morris, "Evolution and the New Age."
[199] Francisco Ayala, "Nothing in Biology Makes Sense Except in the Light of Evolution: Theodosius Dobzhansky, 1900-1975." Journal of Heredity (Volume 68, Number 3, 1977), 9.
[200] Dietrick E. Thomsen, "A Knowing Universe Seeking to be Known." Science News (Volume 123, February 19, 1983), 124. Quoting George Wald.

Pantheistic evolutionism may be exceeding the other two versions in popularity today, as atheism is often viewed as intellectually and emotionally bankrupt while theistic evolutionism is too closely tied to the Church, which is wholly rejected by many people today. This mystical version of evolutionism may be where biblical creationists need to become more knowledgeable in the future.

Science Fiction & Evolutionism

Many of the popular science and science fiction writers of the late nineteenth and early twentieth centuries were atheists or, at the very least, agnostics.[201] They spread their philosophy through their writings, which the public could not seem to get enough of during this time. Besides molecules-to-man evolutionism, these writers were also proponents of the extraterrestrial hypothesis, claiming that life had evolved not only here on the earth but throughout the universe as well. This hypothesis teaches that human beings are really nothing special, just another evolutionary accident in the cosmos.

Among the great science fiction writers who promoted evolutionism in their books, three stand out: H.G. Wells, Isaac Asimov, and Carl Sagan. Herbert George (H.G.) Wells (1866-1946) of England is considered by many to be the father of modern science fiction. Wells studied biology under Thomas Huxley, known affectionately among evolutionists as 'Darwin's Bulldog,' at the Normal School of Science in London. It was said that Wells greatly admired Huxley, and during this time Wells went from being a Judeo-Christian moralist to an anti-establishment socialist. Darwinism is credited for having the greatest impact on his descent into atheism. Wells was a strong advocate of eugenics. In fact, he was considered by many to be more extreme in his wish for genetic control through extermination of the weak, mentally handicapped, alcoholic-impaired, and 'racially inferior' than was Hitler! Wells is also considered to be the first person of influence to advocate the extraterrestrial hypothesis. Amazingly, one of his good friends was the English apologist G.K. Chesterton, one of the great Christian defenders of his time. The two men enjoyed debating each other on many occasions.

Isaac Asimov (1920-1992) was a Russian-born American who was an unbelievably prolific writer, with over 500 books to his credit on subjects as diverse as anatomy and physiology, astronomy, biblical studies, biology, chemistry, geography, Greek mythology, history, humor, mathematics, physics, and, of course, science fiction. Evolutionism was well-represented in his writings, as was the extraterrestrial hypothesis. He signed the Humanist Manifesto II and was the 1984 Humanist of the Year.

The American astronomer Carl Sagan (1934-1996) claimed to be an agnostic, although he is famous for his very atheistic statement that the cosmos is all that

---

[201] Carl Sagan was one of the few popular science and science fiction writers who claimed to be an agnostic, but his now-famous statement from the television series *Cosmos*, "the cosmos is all that is or was or ever will be" describes a clear-cut atheism far more than it does agnosticism.

is, or ever was, or ever will be. Many of his Jewish ancestors preferred Genesis 1:1 over his evolutionary worldview, however. Sagan's most popular quote was not exactly a new idea. The fifth century BC Greek philosopher Heraclitus wrote that, "This world, which is the same for all, no one of gods or men has made. But it always was, is, and will be…"[202] It turns out that Sagan's famous line was not that original after all. There really is 'nothing new under the sun' (Ecclesiastes 1:9), is there? Sagan played a leading role in the NASA space program, briefing all Apollo astronauts before their lunar flights, as well as leading experiments in preparation for the Mariner, Viking, and Galileo planetary expeditions. Sagan also provided the biggest boost in public popularity that SETI has ever enjoyed. In his science fiction book-turned-movie *Contact* he explored the issues of extraterrestrial life and atheism versus theism. Sadly, Sagan died young from an aggressive bone marrow cancer. I think he would have been an interesting man to talk to, despite our worldview differences. (Or, maybe because of them.)

The Rise of the Intelligent Design Movement

The decade of the nineties saw the rise of the Intelligent Design Movement (ID), which aims to critique Darwinism strictly on scientific grounds with no mention of the God of the Bible. ID maintains that nature exhibits patterns that can only be explained as the products of an intentional design rather than undirected, random processes of nature. ID may be summarized in the popular phrase, 'An intelligent design demands an intelligent Designer.' Proponents of ID stop short of declaring that the intelligent Designer is none other than the God of the Bible, instead being content to simply point out that Darwinism is insufficient to explain the origin and diversity of life in the world today, as well as in the fossil record. ID utilizes a variety of academic disciplines to make a comprehensive case for a Designer – from astrophysics to microbiology, and everything in between. For ID proponents, no stone is left unturned when pointing out the erroneous assumptions that are being made in the naturalistic view of origins.

There were a few key books that helped launch ID. 1985's *Evolution: A Theory in Crisis*, by the Australian biochemist Michael Denton, is considered by many to be the book that established ID. *Of Pandas and People*, by the American biology professor Dean Kenyon, was published in 1989 and is also heavily credited with advancing ID. *Darwin on Trial*, the 1991 book by the American lawyer Phillip Johnson, further fueled interest in ID. Johnson coined the term 'Intelligent Design' around this time.

One of the amazing points about ID is that it involves several scientists and scholars from various fields of study, and from a wide variety of religious

---

[202] Danny Faulkner, "Heraclitus: Original Proponent of the Eternal Universe." https://answersingenesis.org/astronomy/heraclitus-original-proponent-eternal-universe/

backgrounds – or from no religious background at all.[203] Since so many of the scientists and scholars in ID are from diverse backgrounds and are united simply in the belief in an intelligent Designer of some sort, many in recent times have been persuaded to give design theory a second look. Although ID is a positive force in the origins debate, it lacks the evangelistic power that biblical creationism provides as the identity of the Creator is avoided in ID. However, it must be noted that ID was never intended as being another version of creation evangelism. This is a fact that creationists need to bear in mind before they start to criticize ID too harshly. The arguments of ID are a good start in the right direction in many evangelistic encounters, provided one follows up with the Gospel message. ID can be a useful tool for biblical creationists, so we need to treat it as an ally and not an enemy.

Phillip Johnson came to my home church in 2006, speaking on ID both there and at Knox College, a private liberal arts college in my hometown. He made the point that, as important as Genesis 1:1 is, John 1:1 is what really counts. Even though ID proponents are content to make the case against Darwinism without bringing up their personal beliefs, they all have an opinion on who the 'intelligent Designer' really is. Scripturally, it is crystal clear that Jesus Christ – the second member of the triune Godhead – is the Creator:

> In the beginning was the Word, and the Word was with God, and the Word was God. He was with God in the beginning. Through him all things were made; without him nothing was made that has been made.[204]
>
> Yet for us there is but one God, the Father, from whom all things came and for whom we live; and there is but one Lord, Jesus Christ, through whom all things came and through whom we live.[205]
>
> For in him all things were created: things in heaven and on earth, visible and invisible, whether thrones or powers or rulers or authorities; all things have been created through him and for him. He is before all things, and in him all things hold together.[206]

But in these last days he has spoken to us by his Son, whom he appointed heir of all things, and through whom also he made the universe.[207]

We should use the ID resources written by Phillip Johnson, Stephen Meyer, Michael Behe, and others because they are tools that prepare us to better share our faith with skeptics and seekers. However, if you are a Christian believer you

---

[203] For example, Michael Denton claims to be an agnostic with no religious affiliation whatsoever.
[204] John 1:1-3, NIV.
[205] 1 Corinthians 8:6, NIV.
[206] Colossians 1:16-17, NIV.
[207] Hebrews 1:2, NIV.

need to point people in the direction of Christ as both the Creator and the Redeemer. Jesus told his disciples – and, by extension, us – to, "Go into all the world and preach the gospel to all creation."[208] That may involve starting the conversation with creationism, but we cannot conclude it there.

Evolutionism in the Twenty-First Century

Today in the twenty-first century, evolutionism has further eroded the acceptance of biblical creationism, as well as attempting to suppress our inherent knowledge of God. Evolutionists say that religion is just a natural evolution of civilization. In other words, 'God was made in the mind of man,' rather than 'man being created in the image of God.' Evolutionists also maintain that one religion is just as good as another since they are all creations of the imagination anyway.

The problem of suffering and evil continues to be an obstacle to belief in God for many. The events of September 11, 2001 as well as multiple wars, social ills, and natural disasters afflicting millions worldwide further highlight the debate between belief and disbelief in a God who loves and cares for his creation. Evolutionism has continued to erode the belief in the Creator by entrenching itself in academia and the sciences. Few seem to question the theory, even among many in the Church. Evolutionism has permeated the East as well as the West, thereby fortifying disbelief and biblical skepticism worldwide.

Christians today are battling many of the secular-based 'ism's' that have been developing for centuries. These 'ism's' include hedonistic nihilism, religious pluralism, moral therapeutic deism, secular humanism, scientism, and evolutionism. There is a single cure for these 'ism's,' however: Biblical creationism. If one can get the issue of origins correct, everything else will tend to follow in the right direction. If people begin to accept the antithesis to biblical creationism – namely, naturalistic evolutionism – only God knows what kind of crazy ideas can follow. As we already know, this is happening today. A relatively new 'ism' in our current skeptical culture is transgenderism. It seems that these new ideas keep coming at us faster than we can keep up with.

Hedonistic nihilism has become a huge problem in our world today, and it is the result of a godless evolutionary worldview taken to its logical conclusion. Hedonistic nihilism is the philosophical belief that life has no intrinsic meaning or value, so you may as well, "Eat, drink, and be merry, for tomorrow you die." For some people, suicide is a legitimate alternative to living in a world without hope. Sadly, we are seeing more and more of this happening today. For some suicidal people, taking others to the grave with them (think 'mass shootings') is happening far too often. Are these horrific social ills solely because of evolutionism? No, of course not. But godless evolutionary ideas are a contributing factor.

Religious pluralism is the idea that 'all religions lead to God, or to truth.' (For

---

[208] Mark 16:15, NIV.

those who do not believe God exists, truth serves as an acceptable substitute.) Never mind that most of the world's religions cannot agree on the nature of God, or even his existence for that matter. Judaism and Islam hold to the belief in one God in one person (strict monotheism). Christianity, on the other hand, holds to one God in three persons (triune monotheism). Hinduism teaches a combination of pantheistic monism and polytheism – millions of gods who are all manifestations of the god-force who is present in all things, known as Brahman. Confucianism, which is technically more of an ethical philosophy than a religion, is agnostic while Buddhism is functionally atheistic. Finally, the humanistic religions are flat-out atheistic. Yet all these religions are supposedly equally valid paths to God or truth, despite their contradictory doctrines. In a postmodern, post-Christian world, contradiction is not a problem, however. Many religious pluralists consider theistic evolutionism to be true. (Since many religions today accept some form of evolutionism, this is not surprising.)

Moral therapeutic deism is the idea that God exists, but he is generally not needed in our day-to-day lives – that is, until we really need his help. Defining moral therapeutic deism depends upon the source that one consults, but in general this philosophy or approach to daily living recognizes a handful of key points: God is responsible for creating the world – like conventional deists, an appeal to evolutionary processes is common among moral therapeutic deists – yet unlike traditional deism God is believed to watch over humanity. God wants us to be good to each other, as taught in the Bible and most religious texts throughout the world. The main goal of life is to be happy, and to feel good about ourselves and to encourage those around us. God does not need to be particularly involved in our lives, except when he is needed to resolve an especially pressing problem or situation. Then we must fully rely upon God, throwing ourselves at his feet and begging for his mercy and guidance. Most people who abide by moral therapeutic deism tend to be universalists. They typically insist that all good people go to Heaven after this life is over. (Defining 'good' could vary considerably, of course.) I contend that many people – even some professing Christian believers – are moral therapeutic deists at times, or maybe even much of the time in some cases. We can all fall into the trap of moral therapeutic deistic thinking. I know I have been guilty more than a time or two. The old saying, "There are no atheists in foxholes" explains moral therapeutic deism. When the going gets tough, even the most secular among us need a helping hand from above.

Secular humanism is the idea that 'man is the measure of all things.' This saying is attributed to Protagoras of Abdera, who lived nearly 500 years before Christ. Once again, there really is 'nothing new under the sun' (Ecclesiastes 1:9). Since Protagoras' time, has mankind finally learned to place God in the lead role? Hardly. In a world where 'man is the measure of all things,' scientism and evolutionism will reign supreme. If there is no God in the picture, science becomes elevated to scientism – we must have a way of explaining everything, especially origins, and science can be the only avenue for answering the big questions of life in a godless culture. Of course, evolutionism becomes synonymous with science in an atheistic world.

However, on a positive note the first decade of the twenty-first century saw further developments in the ID response to naturalistic evolutionism. In 2001 the American philosopher and mathematician William Dembski formed the International Society for Complexity, Information, and Design, promoting scientific reasons to believe in ID. ID has made itself known in public education as well. In 2004 Ohio adopted educational standards sympathetic to ID, yet many state legislators have not been quite so supportive of the movement.

Creationists to the Rescue!

The efforts of Answers in Genesis have provided biblical creationists – and the public as a whole – with great resources in recent times. The Creation Museum in Petersburg, Kentucky opened in 2007, while the nearby Ark Encounter opened in 2016. Both museums, which are known widely among even the unbelieving world, teach the truth of creation and the Flood as revealed in Genesis. Likewise, the Institute for Creation Research's Discovery Center for Science & Earth History opened in 2019 in Dallas, Texas. Creation Ministries International has made its presence known as well, featuring some of the best creationist authors and speakers worldwide, as well as producing several excellent creation-based films. Last, but certainly not least, is the oldest of the major creation groups, the Creation Research Society. Formed in 1963, they are responsible for much of the cutting-edge research that comes from creation-based science. These four groups – Answers in Genesis, the Institute for Creation Research, Creation Ministries International, and the Creation Research Society – are the major creation groups today.

Besides these four groups, there are countless regional and local ministries that are answering the call to spread the message of biblical creationism. One example of a successful local group is Helmut Welke's Quad Cities Creation Science Association (QCCSA), located on the Illinois-Iowa border and encompassing several cities such as Moline and East Moline on the Illinois side of the Mississippi River, and Davenport and Bettendorf on the Iowa side.

Welke, a retired engineering manager with a Fortune 100 company, and his QCCSA group serve not only the greater Quad Cities area but attract members and interested learners from a large surrounding area as well.[209] Welke has worked with creationists from the major creation science ministries, and he has spoken on creationism not only throughout America but in Germany, Poland, and France as well. His interests include not only all aspects of creation science, but also countering the claims of atheism. Local groups such as QCCSA are really a grassroots movement for creationism that has impacted our culture, and the Church, in a positive way.

---

[209] This is a large metropolitan area with a population close to 400,000 people.

An example of a larger group that is more regional in scope is Midwest Creation Fellowship, which serves the greater Chicagoland area. They are known for offering world-class creationist presentations throughout the Chicago area. I was fortunate enough to have one of their top presenters, Dr. Kenneth Funk, review this book prior to publication. (Dr. Funk is also the Director for Midwest Creation Fellowship-North.) His comments and suggestions helped me tremendously during the time that I was finishing this book.

Not to be outdone by their young earth counterparts, Reasons to Believe and BioLogos have been a huge resource for old earth creationists and theistic evolutionists worldwide. If anything, these vastly different ministries should serve to sharpen our individual apologetic. There is nothing like the claims of a competitor – whether they are fellow believers or otherwise – to sharpen one's case for creation: "Do your best to present yourself to God as one approved, a worker who does not need to be ashamed and who correctly handles the word of truth."[210]

The future appears to be quite interesting in the continuing debate on origins. Sadly, however, the Christian position on origins has found itself divided into several camps, with too much fighting between them. Creationists need to be apologetically prepared to make the case for their view, all the while offering the respectful gentleness that God asks of us (1 Peter 3:15). We need to bring the doctrine of biblical creationism to a world in need of the hope that is found only in the one true God of the Bible.

'There is Nothing New Under the Sun'

As we have seen, evolutionary ideas are more prominent today than ever before, which is hardly surprising in a fallen world that has not yet been regenerated in Christ (Revelation 21:1-4). Yet most people today assume that evolutionism is a modern idea, originating with Charles Darwin less than 200 years ago. Evolutionism is an ancient idea, going back at least to the time of the Greeks and Babylonians, and likely even earlier. After thousands of years, the battle for origins has not let up. As Solomon wrote three thousand years ago, "there is nothing new under the sun."[211]

We have traced the history of the origins debate from the ancient world to the present. If you are like I was several years ago, you are amazed at how long the battle between evolutionism and creationism has been raging on. We must now turn our attention to the evidence for God and biblical creation: It is not enough to simply say that the battle is long, we must understand why creationism is to be believed. Most books which examine the history behind the origins debate – and those books are few and far between – rarely address the evidence for creationism. I want to not only examine the evidence for God and biblical creation, but also address other primary skeptical objections as well. Addressing

---

[210] 2 Timothy 2:15, NIV.
[211] Ecclesiastes 1:9, NIV.

every skeptical objection would be a monumental task, and that is beyond the scope of this work. I will speak to the two objections that I have found to be most common among the unbelieving world: The problem of suffering and evil, and salvation in Christ alone. For those who are interested in exploring skeptical objections beyond just these two, I highly recommend both *The Problem of God: Answering a Skeptic's Challenges to Christianity*, by Mark Clark and Ray Comfort's *Faith is for Weak People: Responding to the Top 20 Objections to the Gospel*. But first, let us look at the 'Case for a Creator.'

CHAPTER FIVE

The Case for a Creator

I have spent the last four chapters describing the history behind the origins debate, and it is clear by now that I have both feet firmly planted in the creationist camp. (Big shocker!) Some of you readers – those who are skeptical, and even those of you who favor theistic evolution – probably think I am a bit of a jerk. I can hear it now: "This guy must really hate evolutionists!" However, nothing could be further from the truth. Evolutionists have a voice in the ongoing origins debate, and I have many friends who are evolutionists – and I cherish them greatly. (Even those of you who subscribe to the atheistic version of evolutionism. Seriously.) My problem with evolutionism is not personal in any way, rather it is based on theological and scientific problems with the theory. In basketball, coaches have been known to shout to their players, "play the ball, not the man." That is how the origins debate works for me. My beliefs are founded upon the evidence, or lack thereof, and not how much I like a certain representative of each camp. (Interestingly, I really like the atheistic evolutionist Michael Shermer, while there are more than a few creationists who I find extremely challenging due to their closed-mindedness. Trust me, this issue is never a personal thing for me.) So, let me take a chapter to explain why I am a creationist – and why I believe that you, too, should at least strongly consider it.

The case against evolutionism that is provided in this chapter refutes the idea of a random, chance universe and an accidental origin of life here on the earth. It has often been said that we should consider the evidence for creationism in a comprehensive manner, as no single line of evidence forms a water-tight case for creationism on its own. I beg to differ, however, as some of these evidence's for creationism are quite convincing on their own. Putting them together into a comprehensive 'Case for Creationism' only makes it that much more convincing.

Having said that, however, it must be acknowledged that evidence may be unconvincing for many skeptics. The unbeliever, just like the believer, inherently knows that God exists; the evidence for creationism is plainly before us. As beings made in God's image (Genesis 1:27), we were hardwired to know that God and the realm of eternity exists (Ecclesiastes 3:11). Additionally, we can easily see God through creation (Psalm 19:1; Romans 1:20), our moral conscience (Romans 2:14-15), and human history (Acts 17:26-28). Nonetheless, unbelievers suppress this knowledge (Romans 1:18-32; 1 Corinthians 2:14; 2 Corinthians 4:4). Therefore, the various lines of evidence that creationists are quick to point out may not persuade a skeptic. Regardless of how convincing the science, history, and logic for creationism may be – and it is powerful – the unbeliever may simply ignore it, or even attempt to explain it away. However, there are times when evidence is of great value. The evidence for God and creation worked for me when I was searching for truth, so it can work for other people as well. But keep in mind that no matter how convincing your apologetic might be, some skeptics will simply not open their minds to the evidence. Our job, as apologists, is to simply present the evidence for creation. God will do the convincing. (Make that 'convicting.') As Paul wrote to the struggling churches in Corinth, "I planted the seed, Apollos watered it, but God has been making it grow. So neither the one who plants nor the one who waters is anything, but only God, who makes things grow."[212]

When considering the big questions of life, people tend to think in one of two ways: Naturally, or supernaturally. Those who see the world through natural lenses are convinced that there is a scientific explanation for everything. (This is the belief known as 'scientism,' which we explored extensively near the end of chapter three.) For these folks, there is no reason to invoke God as the reason for anything. We are just here, and probably for no reason. And when we die, we are gone forever. In their worldview, science will eventually explain almost everything, so forget about adding God to the equation. Although these skeptics know they cannot explain everything, they have 'faith' that someday the learned elite among us will figure it all out. (Or at least enough of it to conclude that God is not necessary.) Therefore, the 'whole God thing' is nothing more than a waste of time for these folks.

On the other hand, people like me see the fingerprints of God everywhere we look. For us, the universe screams design, and we know that these earthly lives are not all there is. However, we must be careful to first look for a natural explanation for things. There is no reason to automatically appeal to a supernatural explanation for something without first seeking a natural (scientific) reason. This was the approach I employed as a seeker of truth, during the time that I moved from being a skeptic to becoming a believer. I became aware that evolutionism was sorely lacking in evidence. Once evolutionism was seen for what it is – a failed hypothesis – the only real choice left for me was to believe in the omnipotent Creator who brought everything into existence. (I did consider both pantheism and deism, but neither of those worldviews could stand up to scrutiny. I outline this search for the truth in my first book, *Worldviews in Collision*.) We must learn to simultaneously see the natural and the supernatural, and we must let common sense dictate which is the likely explanation for something. For me, it became clear that the ultimate source of the universe, and everything in it, is God.

Also keep in mind that the case against evolutionism is not just focused on biology. That is only one aspect of evolutionism. In the broad sense of the term, evolutionism consists of cosmological, geological, and biological aspects, so we will examine what Scripture and science has to say regarding all three of these areas.

In this chapter we will look at just three lines of evidence that unmistakably point us in the direction of the Creator who is revealed in the pages of Scripture. These three lines of evidence are (1) the universe had a beginning, (2) divine

---

[212] 1 Corinthians 3:6-7, NIV.

design is everywhere, both in the physical universe and in life itself, and (3) the cohesiveness between science and Scripture.

A Universe with a Beginning

Scripture begins exactly where it should: "In the beginning, God created the heavens and the earth."[213] Ancient cultures, with the notable exception of the Hebrews, believed the universe to be eternal, or without a beginning. Only in the first half of the twentieth century did scientists finally begin to recognize that time, space, and matter-energy came into existence at a specific point in the distant past. Even Albert Einstein found the concept of a beginning for the universe difficult to accept when first presented with the evidence, as he originally held to an atheistic worldview replete with eternal matter.[214] Yet he eventually accepted the fact that the universe is not eternal, a concept that his ancient Hebrew ancestors understood fully.

The 'when' of creation is a fascinating topic as well, but it pales in comparison to the fact that the universe had a beginning and is not eternal. (We will look at the 'when' of creation in the next chapter.) The fact that the universe had a beginning is powerful evidence for the Creator.

The History of the Beginning

The Hebrew word for 'created' is *bara,* which some scholars insist means 'creation from nothing.' (We already examined the original language of Genesis, albeit briefly, in chapter two.) *Bara* may signify that something has been brought into existence when there formerly was nothing at all. Other scholars believe that *bara* refers to any creative act of God, but not necessarily creation from nothing. (It could include creation out of pre-existing matter as well.) Whether or not this Hebrew word is limited to creation from nothing, this concept is discussed in the New Testament by the writer of Hebrews: "By faith we understand that the universe was created by the word of God, so that what is seen was not made out of things that are visible."[215] We can say with confidence that the universe was ultimately created from nothing: There was a time before the universe began – before "in the beginning" took place – and then the universe was. Yet at the time that these verses were being written, the pagan nations surrounding the Hebrews were teaching some form of cosmic evolutionism from pre-existing (eternal) matter.

Exactly why the ancient civilizations held to the belief in an eternal universe is not always clear. Danny Faulkner, a creationist astronomer with Answers in Genesis, explains why the Greeks may have viewed the universe as eternal:

---

[213] Genesis 1:1, ESV.
[214] Although Einstein would later become a pantheist.
[215] Hebrews 11:3, ESV.

There likely are two reasons for this. First, the ancient Greek gods were very limited: not much more than super-men. Therefore, their gods were incapable of fiat creation [creation from nothing]. Second, from their study of how the world around them operated, the ancient Greeks couldn't see any possibility that the world could have created itself. If no one created the universe and the universe didn't create itself, then the only other possibility is that the universe was not created, and hence must have always existed.[216]

It was not just the Greeks who reasoned this way, but many other ancient cultures as well. Concerning the nations of the ancient world, if their gods were not big enough to create everything from nothing, and they understood that the universe could not create itself, then the only option remaining is a universe that has always existed. The only three options are (1) someone made the universe, (2) the universe made itself, or (3) it has always existed. As we will see, the second and third options are not reasonable.

However, there may be another explanation as well. In Hinduism, Brahman is the 'Ultimate Reality,' a god-force that is uncreated and eternal. Everything in the universe emanates or issues forth from Brahman. Therefore, since Brahman is eternal, the universe – which issues forth from Brahman – would also be eternal. Since Pythagoras is credited with introducing Hindu Brahmin teachings to ancient Greece, there may also have been a pantheistic influence in the Greek isles as well. For those in Greece who were polytheistic (many gods), Dr. Faulkner's explanation may best explain the belief in an eternal universe. On the other hand, those who were more pantheistic, like Pythagoras, may have aligned more closely with the latter explanation (god-force emanation). It is difficult to say for sure if this philosophical duality existed in ancient Greece, but it could serve as an explanation.

Regardless of why ancient people accepted an eternal universe, it is most incredible that this belief persisted all the way into the twentieth century. Amazingly, many people today still believe that the universe has always existed. Yet the Bible begins with these very words: "In the beginning…"[217] Even some of the early Christian thinkers accepted the idea that the universe had always existed, but that God specially created the earth at a finite point in the past. This is a strange idea, to say the least, but this erroneous notion does highlight the fact that Greco-Roman cosmology did wield a strong influence upon Christian theology at one time. Dr. Faulkner addresses this idea of an uncreated universe combined with a created earth:

---

[216] Faulkner.
[217] Genesis 1:1, ESV.

Perhaps many of them [early Christian theologians] thought the creation account of Genesis 1 referred to the creation of the earth and things on earth. None of them seemed to have directly opined on this question. At any rate once the secular scientists settled on dismissing the biblical account, the eternality of the universe was taken as a given by many astronomers and physicists until the rapid ascendency of the big bang model a half century ago."[218]

Perhaps the main point is this: Even the early Christian scholars were influenced by the pagan world. That very same thing can happen to us today if we neglect to separate scriptural truths from non-biblical thinking.

Dr. Faulkner sheds further light on this surprising insight:

For a long time, I had been aware that many ancient Greeks thought the universe was eternal. As with so many other ancient Greek ideas, the belief that the universe had no beginning became a common theme of Western thought. Even as Christianity came to dominate the West by the fourth century, belief that the universe was eternal persisted. How did people in the West square this with the "In the beginning" of Genesis 1:1?[219]

The Israeli scientist Gerald Schroeder, an intelligent design proponent, also notes how influential the ancient belief in an eternal universe was even in modern times:

In 1959, a survey was taken of leading American scientists. Among the many questions asked was, "What is your estimate of the age of the universe?" Now, in 1959, astronomy was popular, but cosmology – the deep physics of understanding the universe – was just developing. The response to that survey was recently republished in Scientific American – the most widely read science journal in the world. Two-thirds of the scientists gave the same answer. The answer that two-thirds – an overwhelming majority – of the scientists gave was, "Beginning? There was no beginning. Aristotle and Plato taught us 2,400 years ago that the universe is eternal. Oh, we know the Bible says, "In the beginning." That's a nice story; it helps kids go to bed at night. But we sophisticates know better. There was no beginning.[220]

As late as 1959 the influence of Plato and Aristotle still held sway in scientific

---

[218] Faulkner.

[219] Ibid.

[220] Gerald Schroeder, "The Age of the Universe." http://geraldschroeder.com/AgeUniverse.aspx

thinking. This shows us two things: One, how we must be careful to let Scripture be our constant guide, and two, just how rare paradigm shifts can be. A paradigm is a perspective or set of ideas that governs how we look at something. In some ways it closely resembles the concept of worldview. Before Copernicus in the sixteenth century, the paradigm of the solar system was geo-centric ('earth-centered'), yet when Copernicus and others in his time and shortly after demonstrated that the sun is at the center of the solar system and not the earth (helio-centric or 'sun-centered'), the reigning paradigm shifted. As the quote from Dr. Schroeder reveals, it was only in relatively recent history that the paradigm shifted from Greco-Roman cosmology (eternal universe) to secular 'Big Bang' cosmology (self-created universe).

Cause-and-Effect: The Cosmological Argument Explained

Besides the scriptural refutation, the principle of cause-and-effect dispels the notion of an eternal universe. In its most basic form, the evidence from cause-and-effect for a universe with a beginning is summarized in three steps: (1) Whatever begins to exist must have a cause. (2) The universe began to exist. (3) Therefore, the universe must have a cause.

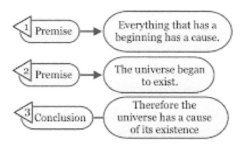

There is one other point of great importance: In cause-and-effect relationships, the cause always transcends (surpasses or goes beyond) the effect. Whatever caused the universe to exist, it transcends the entirety of the cosmos. Think about that for a moment: What could possibly transcend the universe itself?

Throughout much of history, the belief in an eternal universe was common. But if the universe had no beginning, then God is 'out of a job,' so to speak. This explains why the 'battle over the beginning' is of utmost importance in the atheism-theism debate. Although we will explore both the science and Scripture behind the beginning of the universe in a moment, for now we can say with great confidence that the universe did, in fact, have a beginning. After all, the first three words of Scripture confirm this fact, and even the most secular of scientists today are convinced of it. Now let us explore this evidence for a universe with a beginning in more detail.

Was the beginning of the universe caused or uncaused? Once again, according to the principle of cause-and-effect everything which has a beginning has a cause. Since the universe had a beginning, it must have had a cause. The claim that the beginning of the universe was uncaused would mean that the universe sprang into existence out of nothing, and for no reason, but it would be absurd to think that this happened. A basic rule of logic is that nothing never produces something.

Did the cause of the universe come from within the universe itself, or from outside of the universe? The cause of the universe cannot lie within the universe itself, for in cause-and-effect relationships the cause always transcends, or lies outside of, the effect. Therefore, the cause of the universe had to come from outside of the universe. This means that the cause of the universe is supernatural, which literally means 'beyond nature' or beyond time, space, and matter-energy. If the cause of the universe originated from within the universe, and the universe had a beginning, this would mean that the cause of the universe sprang into existence at the exact same time that the universe did. Of course, this is nonsensical, as causes always precede their effects.

Does this cause from outside of the universe need a cause itself? Either the cause of the universe was caused to exist by something else, or the cause of the universe does not depend upon anything else for its existence. Although the principle of cause-and-effect states that everything that has a beginning has a cause, what if the 'first cause' of the universe was itself eternal, or without a beginning? We know that the universe had a cause, because it had a beginning, but that does not mean that the cause of the universe had to have a beginning. If the cause of the universe is not caused by something else – in other words, if this cause is eternal – then there are serious theological implications involved. (Think 'God.') The Bible describes the one true God who brought the universe into existence as being eternal, or without beginning:

> Lord, you have been our dwelling place throughout all generations. Before the mountains were born or you brought forth the whole world, from everlasting to everlasting you are God.[221]
> "I am the Alpha and the Omega," says the Lord God, "who is, and who was, and who is to come, the Almighty."[222]

Since the universe had a beginning it had a cause, and by extension everything within the universe had a cause as well. But the cause of the universe is itself uncaused. Countless people throughout history have recognized the identity of this First Cause as being the God who reveals himself in Scripture.

[221] Psalm 90:1-2, NIV.
[222] Revelation 1:8, NIV.

Table 5.1

Cosmological Argument

THE UNIVERSE HAD A BEGINNING.

THE BEGINNING OF THE UNIVERSE WAS CAUSED.

THE CAUSE OF THE UNIVERSE CAME FROM OUTSIDE THE UNIVERSE ITSELF.

THE CAUSE OF THE UNIVERSE IS UNCAUSED

(ETERNAL OR WITHOUT A BEGINNING ITSELF).

Science Points to a Beginning

At this point, an examination of four of the most important scientific pointers to the beginning of the universe is in order. The First Law of Thermodynamics, which describes the conservation of mass and energy, states that the totality of matter and energy can be neither created nor destroyed naturally. However, one may be converted into the other as demonstrated by Einstein's Theory of Special Relativity, the famous $E = mc^2$ equation. Since matter-energy may not be created naturally, yet matter-energy exists, it therefore had to originate somehow. It is logical to maintain, therefore, that matter-energy was created supernaturally, 'in the beginning,' before this law was put into place. No new matter-energy is being created, which agrees with the words of Scripture: "Thus the heavens and the earth were completed in all their vast array. By the seventh day God had finished the work he had been doing; so on the seventh day he rested from all his work."[223]

The Second Law of Thermodynamics informs us that the contents of the universe are becoming less ordered over time, and the amount of usable energy in the universe is constantly decreasing. Although the first law tells us that matter-energy cannot be created naturally, the second law tells us that matter-energy is degrading over time. In short, the universe is wearing down, a phenomenon known as entropy. Therefore, it is impossible for the universe to be eternal, as it could not have been dissipating or 'winding down' forever. The Second Law of Thermodynamics is accurately described in Scripture (Psalm 102:25-26; Isaiah 34:4; 51:6), once again confirming the positive relationship between science and the Bible.

The expansion of the universe, as described in the Old Testament (Psalm 104:2; Job 9:8; Isaiah 40:22; 42:5; 44:24; 45:12; 48:13; 51:13; Jeremiah 10:12; 51:15; Zechariah 12:1), has been scientifically confirmed through the phenomenon of red shift. Red shift is attributed in large part to the work of the astronomer Edwin Hubble in the early part of the twentieth century, and merely confirmed what the biblical writers already knew millennia earlier. Using the

---

[223] Genesis 2:1-2, NIV.

most powerful telescope in the world, Hubble observed that the universe is expanding in all directions. He discovered that a galaxy's velocity is proportional to its distance: Galaxies that are twice as far from the earth are moving away twice as fast. Since the universe is expanding in every observable direction, some people believe that creation began from a near-infinitesimal starting point. (Perhaps a point even smaller than the period at the end of this sentence.) However, we cannot say with certainty that the universe began in this way. Some recent creationists could be uncomfortable with this idea, as a near-infinitesimal starting point for space seems to align with current 'Big Bang' thinking. However, a spatially near-infinitesimal starting point does not necessarily mean that the Big Bang model is correct. Even in the recent creationist model, God could have begun the expanding universe from either a near-infinitesimal point or from an initially large structure. We cannot say with any degree of certainty. What we can say with supreme confidence, however, is that God initiated the creation event: Time, space, and matter-energy became a reality when God willed the cosmos into existence.

In 1965, two scientists from Bell Labs, Arno Penzias and Robert Wilson, attempted to detect microwaves from outer space, and inadvertently discovered an electronically detected 'noise' of extraterrestrial origin. This was five years after SETI – the Search for Extraterrestrial Intelligence – began under the initial name Project OZMA, so as would be expected there was much excitement generated at first. However, the noise did not seem to emanate from one location in the universe, but rather came from all directions simultaneously. Therefore, the noise that they heard was of a natural origin, as opposed to being attributed to advanced, thinking beings from elsewhere in the universe. Penzias and Wilson went on to discover that this detectable noise was radiation which had been left over from the creation event itself – sort of a 'fossil remnant' of the universe's beginning. The discovery of cosmic background radiation lent strong support to the idea of a universe with a beginning. More recently, NASA's COBE satellite was able to detect cosmic microwaves emanating from the furthest reaches of the known universe. The uniformity of these microwaves demonstrated the homogeneity ('same condition throughout') of the early universe, shortly after the initial moment of creation. As the universe cooled, yet maintained its expansion, small fluctuations began to form from temperature differences. These temperature differences verified calculations that had been performed in the attempt to demonstrate the hypothesized cooling and development of the universe just fractions of a second after its creation. Once again, although this may seem heretical to some recent creationists – as it seems to describe secular Big Bang cosmology – it does not rule out Genesis creationism in the recent past.

Table 5.2

Scientific Evidence for a Beginning

FIRST LAW OF THERMODYNAMICS

SECOND LAW OF THERMODYNAMICS

EXPANSION OF THE UNIVERSE (RED SHIFT)

COSMIC BACKGROUND RADIATION

Answering the Skeptics

Due to the formidableness of cause-and-effect as evidence for the Creator, skeptics are often quick to attack this line of evidence. In this section we will examine three common skeptical arguments against the cosmological argument. The first argument concerns a misunderstanding of God's nature. It is common for skeptics to ask the question, "If God made the universe, then who or what made God?" In their minds, if everything requires a cause, then something or someone must be responsible for God's existence. However, as already pointed out the principle of cause-and-effect does not state that everything has a cause, but rather everything which begins to exist has a cause. If God did not begin to exist, but rather is eternal – a point which is declared throughout Scripture (Psalm 90:1-2; 1 Timothy 1:17; Hebrews 13:8) – then God himself does not need a cause. In fact, God alone is eternal or without beginning. The entirety of the universe is caused, but God is not. God is eternal, infinite, immaterial, and immutable – he has no beginning, no limits, no physical body, and he never changes. God is uncaused, a concept which is impossible to wrap our finite, limited minds around. God, through the words of the prophet Isaiah, declares his infinite character:

> "For my thoughts are not your thoughts, neither are your ways
> my ways," declares the LORD. "As the heavens are higher than
> the earth, so are my ways higher than your ways and my thoughts
> than your thoughts."[224]

The belief that God himself had to have a cause is based upon a faulty understanding of both the principle of cause-and-effect and the nature of God. It is a common misunderstanding that skeptics would do well to abandon.

The second argument against the beginning of the universe involves the branch of science known as 'quantum physics.' Some skeptics have claimed that nothing can produce something, and they appeal to quantum physics as the 'hero'

---

[224] Isaiah 55:8-9, NIV.

which makes that possible:

> According to contemporary quantum physics, virtual particles can pop into existence in what is called the 'quantum vacuum.' They are called 'virtual particles' because they cannot be observed directly, although their indirect effects can be measured. Because the quantum vacuum doesn't appear to be an object – according to our ordinary notion of object – it is sometimes said that virtual particles literally come into being out of nothing.[225]

However, we must ask the question, "Where did this quantum vacuum supposedly come from?" No matter what we may think of this quantum vacuum, it is most definitely something, even if non-physical (immaterial) in nature. However, even proponents of the quantum vacuum idea say that it gives every indication of being a physical object with a complicated structure, and the term 'structure' itself further implies that it is physical in nature. "[The quantum vacuum] is simply the lowest energy state of the quantum field, and the quantum field is a physical object with a very complicated structure, a structure that is specified by a set of equations that contain a variety of apparently arbitrary numbers."[226]

When the quantum vacuum is no longer in its lowest energy state, particles do appear to pop into existence, but it turns out there are physical events taking place at the atomic or even subatomic level which merely give the appearance of the creation of these particles from out of nothing. True 'nothingness' has no physical or non-physical properties: 'Nothingness' cannot be studied by science, because there is absolutely nothing to study! Quantum physics, although fascinating, is an area of the physical sciences that still leaves many questions unanswered.[227]

The third argument against the beginning of the universe involves the so-called 'oscillating theory.' This theory was devised as a way around the problem of a universe with a beginning, which clearly contains theological implications. British physicist John Gribbin is refreshingly honest about this: "The biggest problem with the Big Bang theory of the origin of the Universe [universe with a beginning] is philosophical – perhaps even theological – what was there before the bang? This problem alone was sufficient to give a great initial impetus to the Steady State theory [universe is eternal]."[228]

Gribbin admits that the theological implications of a beginning point for the

---

[225] James R. Beebe, "The Kalam Cosmological Argument for the Existence of God." http://apollos.squarespace.com/cosmological-argument/

[226] Ibid.

[227] There are some who claim that, once quantum physics is better understood, there will be spiritual or theological implications involved. Quantum physics may very well point us in the direction of God.

[228] Hugh Ross, *The Fingerprint of God* (New Kensington, PA: Whitaker House, 1989), 97.

universe are problematic for many people, as cause-and-effect can rule out naturalism. Gribbin continues: "…but with that [Steady State] theory now sadly in conflict with the observations, the best way round this initial difficulty is provided by a model in which the universe expands from a singularity, collapses back again, and repeats the cycle indefinitely."[229] This idea is nothing new, which is no surprise (Ecclesiastes 1:9). Hugh Ross, an old earth creationist with a doctorate in astrophysics, relates that both Hindu philosophers and Roman naturalists in the ancient world devised this idea long ago: "The implications of such a prospect [universe with a beginning] so disturbed a number of scientists that they turned to the model of early Hindu teachers and Roman atheistic philosophers – the oscillating universe."[230] As you may recall, we already saw this in chapter one. As mentioned there as well as earlier in this chapter, Pythagoras was likely the Greek philosopher who introduced Hindu Brahmin cosmology to the Greeks.

Proponents of the oscillating theory claim that the universe is eternal, but only appears to have had a beginning because the universe is undergoing an eternal or never-ending series of expansions and contractions. According to these theorists, at a certain point the universe can no longer expand, at which time it begins to contract back to a near-infinitesimal point, only to begin the expansion phase all over again. From our perspective as observers of the universe – utilizing the best scientific tools at our disposal – the universe appears to be expanding for the first and only time. However, proponents of this theory claim that the universe is only expanding during our particular 'bounce,' the term for each individual cycle of expansion-contraction. Perhaps in previous bounces (universes) there were different life forms unlike anything we know of today, and maybe in future bounces there will be new life forms that are unimaginable to us, each following a unique evolutionary pathway. According to oscillating theorists, this current bounce is marked by our existence – in other words, this universe is 'our time' – although many people hold to the idea that human beings are only one of countless life forms throughout our universe. For these extraterrestrial believers, we live in a *Star Trek* universe.

However, the oscillating theory is based upon pure speculation. Furthermore, science does not support the idea. The theory is a convenient way to avoid a Beginner (God), and it is besieged by a host of scientific problems. Despite scientific obstacles, the philosophical motivation to avoid a beginning for the universe – and hence the necessity of a Beginner – is powerful for some people. Surprisingly, in 1973 two Soviet scientists, Igor Novikov and Yakob Zel'dovich, provided scientific evidence against the possibility of oscillation. They pointed out that a uniform isotropic compression, which has been hypothesized as the means of contracting the universe from the point of maximum expansion, "becomes violently unstable near the end of the collapse phase, and the

---

[229] Ibid.
[230] Ibid.

collapsing medium breaks up into fragments."[231] In other words, oscillation cannot work. They went on to state that "oscillation provides no escape from an ultimate beginning in the finite past."[232] As Soviet scientists who were supposed to be devoted to the propagation of Marxist ideology – which is atheistic – it was refreshingly honest of these two men to admit the faulty scientific reasoning behind this theory.

The evidence against the oscillating theory is devastating. First, the observed density of the universe appears to be only half of what would be necessary to force a contraction phase. Second, there is no mechanism or force known that could trigger a contraction phase. Third, isotropic compression becomes violently unstable near the end of the contraction phase, thereby ruling this out as a possible explanation for the 'bounce back' (expansion). Finally, the observable entropy in the universe is such that a bounce becomes impossible after a certain number of bounces, ruling out an infinite number of expansion-contraction cycles. (We end up with a beginning point after all, which is what proponents of the theory were trying to avoid in the first place.) Despite this elaborate attempt to refute a beginning for the universe, the oscillating theory falls flat because of its scientific errors.

The Bible opens with the words, "In the beginning God created the heavens and the earth."[233] Therefore, right at the very beginning of Scripture is the proclamation that God created everything from nothing. The entirety of the universe was brought into existence at a definite point in the distant past. As opposed to many ancient thinkers who believed that either God or the gods formed the world out of eternally pre-existing matter, this proclamation by Moses was radical in its day – and is, in the minds of many, clear evidence that God is behind the words of Scripture.

A universe with a beginning is powerful evidence for the God of the Bible, refuting atheism, pantheism, and deism. (These are the other major worldviews besides Christian theism, which I refuted in my first book, *Worldviews in Collision*.) Although many creationists avoid the cosmological argument, we should all use this pointer to God in our apologetic approach.

A Precisely Designed Creation

Seeing God through nature is nothing new: Aristotle could see God through creation, although his version of God was not the transcendent God of the Bible. Nonetheless, his 'First Cause' of the universe was worlds apart from the atheism that existed even in his day. A few centuries later, Cicero saw the evidence for God in nature better than anyone else in his time. Of course, the biblical writers wrote about God's majesty reflected through creation better than anyone ever has, and there have been countless people since the time of the early Church who

---

[231] Ibid.
[232] Ibid.
[233] Genesis 1:1, NIV.

have written about God's glory in creation. Names such as Augustine, Aquinas, Pascal, Paley, Chesterton, Lewis, Geisler, and Morris have eloquently described God's 'eternal power and divine nature' (Romans 1:20) better than most. Regardless of where we look in nature, and whether we use a telescope, a microscope, or simply the naked eye, divine design is everywhere.

Design throughout the Universe

The universe is clearly designed, and it is anything but random: "The heavens declare the glory of God, and the sky above proclaims his handiwork."[234] The physical parameters that govern the universe are exactly what are required for intelligent life here on the earth. For instance, the electromagnetic coupling constant binds electrons to protons. If this very precise value were just slightly smaller fewer electrons could be held in place, and if it were just slightly larger electrons would be held together too tightly, not allowing for bonding with other atoms. Either way, life could not exist on the earth. Likewise, the ratio of the mass of an electron to the mass of a proton is incredibly precise, 1:1,836. If this ratio were just slightly smaller or larger – even to the third decimal place – molecules could not form, and once again life could not exist on the earth.

Our sun is the right mass. If it were slightly larger, its brightness and radiation levels would be much too high to support life on the earth. Conversely, if it were slightly smaller it would be too cold and would lack the appropriate radiation levels to support life on the earth. Photosynthesis cannot take place unless the sun is the precise size that it is, and without photosynthesis there is no life. Also, the earth's distance from the sun is critical for a stable water cycle. If the sun is too far away water would freeze, and if the sun is too close water would boil. It turns out that the sun is the exact distance from the earth that it should be, to support life on the earth.

The earth's gravity, axial tilt, rotational period, magnetic field strength, crust thickness, oxygen-to-nitrogen ratio, and levels of carbon dioxide, ozone, and water vapor are exactly where they should be to support life on the earth. These are only a partial list of parameters, all having to be extremely precise individually and in perfect conjunction with each other for life to exist on our planet. We truly live in a 'Goldilocks universe' where everything is 'just right.'

But how 'just right' is just right? The astronomer Hugh Ross notes the extreme accuracy of one parameter in the universe:

> Unless the number of electrons is equivalent to the number of protons to an accuracy of one part in 10^{37}, or better, electromagnetic forces in the universe would have so overcome gravitational forces that galaxies, stars, and planets never would have formed.

---

[234] Psalm 19:1, ESV.

One part in 10^{37} is such an incredibly sensitive balance that it is hard to visualize. The following analogy might help: Cover the entire North American continent in dimes all the way up to the moon, a height of about 239,000 miles. (In comparison, the money to pay for the U.S. federal government debt would cover one square mile less than two feet deep with dimes.) [Keep in mind that Dr. Ross wrote this in 1993.] Next, pile dimes from here to the moon on a billion other continents the same size as North America. Paint one dime red and mix it into the billion piles of dimes. Blindfold a friend and ask him to pick out one dime. The odds that he will pick the red dime are one in 10^{37}. And this is only one of the parameters that is so delicately balanced to allow life to form.[235]

That, my friends, is precise! A few pages later, Dr. Ross lists twenty-six parameters that serve as evidence for the fine-tuning of the universe, and several pages after that he lists an additional thirty-three parameters that demonstrate cosmological necessities for life to exist on the earth.[236] All fifty-nine parameters have a precision like that described in the analogy above. We are not talking about one, or even a few, precise parameters, but dozens. (There are other parameters even beyond the fifty-nine that Dr. Ross notes in his book.) What are the chances that all fifty-nine of these parameters could individually be 'just right' apart from divine design? I argue that, apart from God's creative input there is absolutely no possibility of this happening. Also consider this: Not only do these parameters need to be individually precise, but they must all fit together with a combined precision that at least equals their individual accuracy. Words like 'awesome' or 'amazing' do not do justice to the fine-tuning found in the universe.

How might some skeptics attempt to refute the fine-tuning argument? With the multiverse theory. Maybe you have heard of this idea before. According to this belief, there are an infinite number of universes. (I fail to get my mind around the majesty of our one universe, let alone an infinite number of them.) Skeptics claim that, in an infinite number of universes, at least one universe will consist of all its parameters being perfectly set for life to occur on at least one planet – and that just happens to be the universe that we live in, of course!

There is one major problem with the multiverse theory, however: It is pure speculation. There is no evidence for any universe beyond the one we live in. In fact, we could never be aware of any universe besides our own. There is no conceivable scientific means of allowing us to peer beyond the borders of our universe, to see what lies beyond. If we could, we would see beyond time, space,

[235] Hugh Ross, *The Creator and the Cosmos* (Colorado Springs, CO: NavPress Publishing Group, 1993), 115.
[236] Ibid.

and matter-energy – in other words, we would see into the realm of eternity. We would not see another universe 'next door.' (For that matter, how might these separate universes bound each other? Would they abut perfectly, or would there be some type of space between them? Since the multiverse idea is pure fiction, the answer to this question and all others is limited only by the imagination.)

Dr. Ross quotes the theoretical physicist Tony Rothman, in an article he wrote concerning the anthropic principle:

> The medieval theologian who gazed at the night sky through the eyes of Aristotle and saw angels moving the spheres in harmony has become the modern cosmologist who gazes at the same sky through the eyes of Einstein and sees the hand of God not in angels but in the constants of nature…When confronted with the order and beauty of the universe and the strange coincidences of nature, it's very tempting to take the leap of faith from science into religion. I am sure many physicists want to. I only wish they would admit it.[237]

I, too, wish they would admit it. I really believe that many physicists want to acknowledge the Creator, but they are afraid of what their secular-minded peers will think of them. Maybe if they did express their thoughts and convictions, the unnecessary tension between secular-based science and religion would dissipate at least somewhat.

Design in Living Things

In addition to the extreme craftsmanship seen throughout the physical universe, we also see incredible design in biology. The Law of Biogenesis, the Cambrian explosion, the complexity of the cell, the information contained within the DNA molecule, and the lack of 'missing links' all scream intentional design in nature. Yet so many people today are blind to this evidence (1 Corinthians 2:14; 2 Corinthians 4:4).

The Law of Biogenesis states that life comes only from life, with no exceptions. Since life obviously exists, and life cannot come from non-living material, the first life forms had to be created. This is exactly what Genesis tells us (Genesis 1:11-12, 20-21, 24-27). The idea of the spontaneous generation of life from non-living matter was initially refuted by Francesco Redi in the seventeenth century, and it was later put to rest by Louis Pasteur at essentially the same time that Darwin released *Origin of Species*. Nonetheless, naturalists must still rely upon the belief that life spontaneously arose from non-living matter to remain consistent in their thinking. Even if the evolution of all life forms came from a single cell, how that single cell could arise from non-living material in the first place is an insurmountable problem for evolutionists. The

---

[237] Ibid.

Law of Biogenesis alone is enough to bury the idea of molecules-to-man evolutionism, yet still it persists.

The Cambrian explosion, also known as the 'Big Bang of Biology,' refers to the appearance in the fossil record of most of the major body plans in an extremely short time. Evolutionists claim that 543 million years ago these new life forms appeared abruptly, in only 20 million years or less.[238] This is considered a rapid burst of new life forms that appears to be inconsistent with the gradual pace of naturalistic evolutionary change. Although the idea of long ages that undergirds this belief has long been debated, the broad concept behind the Cambrian explosion points out a major inconsistency in the naturalistic view of origins: The Cambrian explosion demonstrates that life essentially 'sprang from out of nowhere.' Complex animals appear abruptly, without any evolutionary predecessors in the rock layers below the Cambrian. This includes trilobites, with their extremely advanced eye structure. These animals appear to have arisen spontaneously and fully formed, with none of the transitional forms (evolutionary predecessors) called for by naturalistic evolutionism. This is, of course, exactly what Genesis describes (Genesis 1:11-12, 20-21, 24-27).

However, in addition to the age of the earth debate the Cambrian explosion is the cause of another contentious matter between young earth and old earth creationists: Death before Adam's fall. Does the Cambrian layer predate mankind? If the Cambrian layer existed before Adam, then old earth creationists are correct to point out that death did, in fact, occur before Adam. However, recent creationists maintain that Scripture clearly reveals that death was introduced to the world through Adam's sin (Romans 5:12), therefore the animals encased in the Cambrian layer had to perish sometime after Adam's fall. If the Cambrian layer is much more recent than old earth creationists claim, then these fossilized creatures died sometime after Adam's fall. All the Cambrian layer really tells us is that complex animals died and were buried in this layer, with no evolutionary ancestors preceding them in layers lower down. If evolutionism were true, we would expect to find less evolved creatures lower down in the fossil record, but we do not. The lack of human fossils in this layer does not necessarily mean that mankind did not exist on the earth when these creatures were buried, and it does not tell us if they were buried before or after Adam's fall. To argue that death existed before Adam's fall, based upon this single rock layer, may be insinuating too much. The Cambrian stratum, with its highly debatable age assignment, aligns with special creation, not evolutionary progression. (We will further examine the issue of death before Adam's fall in the next chapter.)

Let us now look at the amazing 'building blocks' of life itself, the cell. When

---

[238] Although the evolutionary timescale is debatable, the main point to keep in mind concerning the Cambrian explosion is that life appeared abruptly, when there previously was no life to be found, and all the major plant and animal forms arose in a very brief time. This is exactly what Genesis reveals to us.

Charles Darwin published *Origin of Species* in 1859, cells were believed to be nothing more than simple globs of protoplasm. Today, however, we know that the cell is so extremely complex that it defies a naturalistic origin. The deeper scientists dig into the microscopic structure of the cell, the more complex and interconnected the cellular parts prove to be.

Among creationists the term 'irreducible complexity' gets thrown around a lot, and for good reason. Irreducible complexity is the idea that certain biological systems cannot have evolved by small, successive modifications over time. Instead, they demonstrate that all their many complex parts had to be fully functioning from the beginning of their existence to the present. The evidence for this is based upon the irreducible interconnectedness of the individual parts. For the system to function, all the parts had to be in place from the beginning, or it simply could not work.

Genetics has become a huge ally for creationists. A 'simple' cell evolving into more complex life forms would require added genetic information, which cannot happen naturally as cells can only work with the genetic information that they have at their disposal. The information contained in the DNA of just one cell is mind-boggling, and each created kind of plant or animal had to have the entirety of its DNA present at the beginning of its existence (Genesis 1:11-12, 21, 24-25). A plant or animal can lose genetic information over time but can never gain genetic information, which is what molecules-to-man evolutionism requires. It does not matter whether an evolutionist is coming from an atheistic or a theistic position – the evolutionary process is the common denominator either way. Genetics rules out the slow, gradual evolution of life.

As amazing as the DNA molecule is, the 'software' – which is the information or 'code' within the molecule – points to a Designer. Information always comes from one source: A thinking mind. To believe that genetic information simply evolved by chance is to accept an impossibility that no rational person, if he or she took the time to thoughtfully consider the matter, would embrace. "The amount of information that could be stored in a pinhead's volume of DNA is equivalent to a pile of paperback books 500 times as tall as the distance from Earth to the moon, each with a different, yet specific content."[239] The preeminent atheist Richard Dawkins admits that DNA has "enough information capacity in a single human cell to store the Encyclopedia Britannica, all 30 volumes of it, three or four times over,"[240] and perhaps more amazing still is that a pinhead of DNA can hold 100 million times more information than a 40 gigabyte hard drive. Yet naturalistic evolutionists are adamant that this molecule simply evolved over eons of time through chance, random processes in nature.

Regarding the ability of DNA to accurately reproduce itself, there is an unmistakably high degree of irreducible complexity involved in this process.

---

[239] Grigg, "A Brief History of Design."
[240] Jonathan Sarfati, "DNA: Marvelous Messages or Mostly Mess?" https://creation.com/dna-marvelous-messages-or-mostly-mess

When a DNA molecule undergoes reproduction, the encoded information is precisely copied into a new strand. This copying is far more precise than pure chemistry can account for: Only about one mistake occurs out of ten billion copies. This is due to an 'editing machine' that is part of the DNA molecule itself. This machine proofreads the code, then corrects for errors that might have arisen had there been no editing (error-checking) system in place. This proof-reading machine had to be in place from the beginning, otherwise DNA would not have continued for long. (It would have degraded into useless information relatively quickly.)

The noted philosopher of science Sir Karl Popper admitted that the DNA code cannot be translated without relying upon certain products of its translation. "This constitutes a baffling circle; a really vicious circle, it seems, for any attempt to form a model or theory of the genesis of the genetic code."[241] That admission is boldly honest, for it is incredibly baffling and truly a vicious circle only if one is locked into accepting a purely naturalistic explanation for the origin of DNA. Only when the possibility of a supernatural explanation for the DNA code is considered can we put an abrupt end to this 'mystery.'

However, the reluctance to consider a supernatural explanation for the origin of DNA can lead to some off-the-wall ideas. Sir Francis Crick, one of the co-discoverers of the DNA molecule, was devoted to atheism long before his monumental discovery. Crick was the 'Richard Dawkins' of his generation.[242] In 1961, the year before he and fellow researcher James Watson discovered the DNA molecule, "Crick resigned as a fellow of Churchill College, Cambridge, when it [the college's administration] proposed to build a chapel."[243] Interestingly, Crick's distaste for religion was one of the reasons that led to his discovery of the molecule. He believed that DNA would provide the evidence that life simply evolved by chance, random processes. However, the staggering amount of precisely encoded information is anything but proof for naturalism, and many are convinced that Crick secretly came to the realization that naturalistic evolutionism had nothing to do with its origin.

FRANCIS CRICK

---

[241] Ibid.

[242] Richard Dawkins is currently the preeminent 'atheistic evangelist' in the world today.

[243] Gary Bates, "Designed by Aliens? Discoverers of DNA's Structure Attack Christianity." https://creation.com/designed-by-aliens-crick-watson-atheism-panspermia

Enter directed panspermia. Panspermia is derived from two Greek words: *Pan*, meaning 'all,' and *sperma*, meaning 'seed.' Panspermia maintains that the seeds of all life are to be found throughout the universe. According to this theory, at the time of the so-called 'Big Bang' or shortly thereafter, life developed and was spread throughout the universe via comets and other bodies traveling through space. Crick preferred the idea of directed panspermia, which holds that the seeds (cells and molecules) constituting all of the life forms here on the earth were directed to this planet by an intelligence from beyond our sphere – an intelligence which does not include God, of course! Crick proposed that primordial life was shipped to the earth billions of years ago by an advanced race of extraterrestrial beings. In his view, these extraterrestrial beings had either evolved naturally or had originated through directed panspermia via an even earlier alien civilization. One can see how that could quickly turn into an almost infinite regress of 'alien creators.' Crick acknowledged the futility of his view on origins when he wrote, "Every time I write a paper on the origins of life, I swear I will never write another one, because there is too much speculation running after too few facts..."[244] Had Crick simply accepted the evidence of divine creation, and acknowledged the Creator revealed to us in Scripture, he would never have had the problems that he wrestled with concerning the origin of life. This is merely one example demonstrating just how far a person will go in his or her attempt to deny the obvious evidence of divine design.

Interestingly, for many years' evolutionists believed that over 95 percent of our DNA was 'junk.' This junk DNA was supposedly left over from our evolutionary past, when we supposedly had fins, tails, and other primitive structures. According to evolutionists, this junk DNA is unused and unnecessary today. Evolutionists considered it to be great evidence for molecules-to-man evolutionism, but today we know that there is no such thing as junk DNA. All our DNA is used in some way. What was once considered 'junk' is now known to be extremely important, and is the controlling genes that control other genes, which control other genes, and on and on. To believe this complex set of instructions – this mind-boggling code – could have arisen by chance is sheer madness. All our DNA is used at some point, making the idea of junk DNA both bad science and bad theology.

The more complex an event or structure is, the less likely that it just happened by chance. For example, the presidential faces on Mount Rushmore are too complex to be the result of wind and water erosion. Anyone who is told that random processes in nature could form the artistic design seen on Mount Rushmore would have their intellect insulted. If the faces on Mount Rushmore show signs of having been designed on purpose, then how much more incredible is it to believe that the DNA code – which is unbelievably complex and information-rich – happened apart from intelligent design? Amazingly, the academic and skeptical communities, which overlap far too often, propose that

[244] Ibid.

the structures of nature evolved into their current states of complexity apart from creative design. A proposal such as this must have an ulterior motive behind it, and perhaps even a personality directing it. Evolutionism is truly a 'doctrine of demons' (1 Timothy 4:1).

Designed to Think: The Amazing Human Mind

Let us now move into the area of human thought processes. Human cognition may be defined as that group of mental processes which are concerned with attention, memory, problem-solving, decision-making, and the ability to produce and understand language. Countless neurologists, psychologists, and philosophers believe that human thought can only be reliable if it is designed, rather than being attributable to chance, random processes in nature. Since our thoughts are generally considered to be reliable, it is reasonable to conclude that human thought is intentionally designed.

If human thought were to be merely the result of random biochemical reactions, then we would have no reason to believe that our mental faculties are reliable. No one could really trust their thoughts if they are merely attributable to chance, random processes of biochemistry. Is it possible that naturalistic evolution could in any way account for human thought? Not in my mind!

Some evolutionists claim that thoughts are an emergent property of the brain. Emergent properties arise from the overall functioning of a system, but do not belong to any single part of that system. Therefore, evolutionists who claim emergence as their explanation for the ability to think believe that thoughts are generated by the entirety of the brain and nervous system working together. The individual parts of the brain can do nothing to generate thoughts. However, when all the parts are combined, a 'near-miraculous' process takes place – thoughts occur! Evolutionists believe this despite a lack of scientific evidence for it. Ultimately, chance processes of nature cannot realistically produce such intricately complex phenomena as memories, reasoning abilities, and emotions. Human thought is clearly attributable to intelligent design.

There is no reason to believe that the material brain is solely responsible for immaterial thoughts. Atheistic evolutionists maintain that the universe is purely material in nature, as there is no God or spiritual realm behind it. But thoughts are immaterial in nature: They are real, but do not consist of atoms and molecules (matter). Why would immaterial thoughts even be part of a purely material universe? Is there even a place for the immaterial in a purely material universe?

Most atheists are quick to admit that our thoughts are truly our own, and not simply the result of biochemical reactions in the brain. Yet the logical conclusion of atheism is that human thinking is nothing more than biochemical reactions generated by the brain. In an atheistic universe, thoughts can be nothing more than that. The atheist who believes that their thoughts are truly their own is borrowing heavily from the theistic worldview. My advice to them? Stay in your own lane, or feel free to come over to the side of truth. (I sincerely hope for the latter.)

Either we were created by God, or we evolved by chance. If we evolved by chance, then we can forget about having this whole God and creation conversation, since it is meaningless if we cannot trust our own thoughts. But if we are created by God, then everything changes: We can be confident that the Creator endowed us with the ability to know things with great certainty, and that includes the existence of the Creator himself. The great philosopher Rene Descartes once wrote, "I think, therefore I am," but he should have written, "I think, therefore God is" since human thought is an obvious design feature given to us by the Creator.

The Cohesiveness of Science & Scripture

If it could be demonstrated that Scripture accurately describes scientific facts not just centuries, but even millennia prior to modern confirmation, would that get your attention? It did for me, as a young man seeking after truth. Despite my former skepticism, I arrived at the point in my life where I had no axe to grind with the truth claims of Scripture: I was willing to go wherever the evidence led. If we are seeking after the one worldview that makes sense of the world, we must put truth ahead of a personal agenda.

Let us look at just a few examples from Scripture which demonstrate that the biblical writers knew scientific truths far in advance of modern discoveries. Seven centuries before Christ, the prophet Isaiah wrote that God, "sits above the circle of the earth."[245] Although the earth was believed to be flat by some ancient scholars–examples include Lactantius (third century) and Cosmas Indicopleustes (sixth century)–the Bible has always described our planet as being spherical or circular in shape. Many skeptics today are quick to claim that the Bible, and its followers, have historically taught that the earth is flat, but that is simply not true. The evolutionist Stephen Jay Gould (1941-2002) noted, "There never was a period of 'flat earth darkness' among scholars (regardless of how the public at large may have conceptualized our planet both then and now). Greek knowledge of sphericity never faded, and all major medieval scholars accepted the earth's roundness as an established fact of cosmology."[246] This is an example of a skeptical objection to the Christian faith that is not true, yet like many skeptical objections it tends to 'stick around' in popular thinking. The Bible always had it right, though.

The writer of Job revealed that God "hangs the earth on nothing."[247] Earth is suspended by gravity in space, a fact not accurately described by science until relatively modern times. Some of the ancient cosmologies had Atlas holding the world on his shoulders (Greek), or the earth resting on the backs of giant

---

[245] Isaiah 40:22, ESV.
[246] Stephen Gould, "The Late Birth of a Flat Earth" in *Dinosaur in a Haystack: Reflections in Natural History* (New York, NY: Three Rivers Press, 1997), 38-50.
[247] Job 26:7, NIV.

elephants, themselves standing on the back of a giant turtle (Hindu Brahmin). The law of gravity exerting itself between solar bodies was described in the Old Testament, albeit in a more poetic form than would be found in a science textbook today. Science caught up to this fact at the time of Copernicus, in the sixteenth century.

The prophet Jeremiah wrote that God would make King David's descendants "as countless as the stars in the sky and as measureless as the sand on the seashore."[248] Ancient people believed that the stars were countable, mostly because they had a limited view of the heavens and could only see a small fraction of the number of stars just in our galaxy alone. Ptolemy, around the beginning of the second century AD, was busy cataloguing the stars, which he had numbered to 1,100. In the early twentieth century, scientists with the help of high-powered telescopes came to the realization that the number of stars was in the billions. Still later, scientists came to understand that there are about a billion galaxies, each containing a billion stars. Today we know that if we were to count the stars at the rate of ten per second, it would take 100 trillion years to count them all – and that is just the ones that we think we know about! However, Jeremiah wrote six centuries before Christ that the stars are uncountable. Clearly the ancient Hebrews knew things about the creation that the rest of the world was ignorant about.

Although the hydrological cycle was not fully explained until the eighteenth century, in large part through the efforts of Perrault and Marriotte, Scripture always gave an accurate description of how water cycles through its various stages, albeit in layman's terms: "He draws up the drops of water, which distill as rain to the streams; the clouds pour down their moisture and abundant showers fall on mankind"[249] and, "All streams flow into the sea, yet the sea is never full. To the place the streams come from, there they return again."[250] Similarly, Matthew Fontaine Maury, the father of oceanography, read the following verse that describes 'pathways in the sea' or what we now call ocean currents: "This is what the LORD says – he who made a way through the sea, a path through the mighty waters."[251] Taking that verse literally, Maury searched the oceans and discovered – and then mapped – major currents that have been used as routes for sea travel ever since.

Utilizing satellite technology, we now recognize the existence of global wind patterns, yet the Bible discussed the existence of these patterns nearly three millennia before our time: "The wind blows to the south and turns to the north; round and round it goes, ever returning on its course."[252] Today, weather experts have confirmed the accuracy of Scripture regarding these wind patterns. Also,

---

[248] Jeremiah 33:22, NIV.
[249] Job 36:27-28, NIV.
[250] Ecclesiastes 1:7, NIV.
[251] Isaiah 43:16, NIV.
[252] Ecclesiastes 1:6, NIV.

agriculturalists today recognize the importance of allowing land to 'rest' every seven or so years, to allow the land to replenish itself. The ancient Hebrew's clearly knew the value of this practice long before our time, however: "But in the seventh year the land is to have a year of sabbath rest, a sabbath to the LORD. Do not sow your fields or prune your vineyards. Do not reap what grows of itself or harvest the grapes of your untended vines. The land is to have a year of rest."[253] Interestingly, this practice mirrors the seven-day creation week, in which God 'worked' (created) for six days and then 'rested' on the seventh day.

The dimensions of Noah's ark turn out to be the optimum construction for a barge-type vessel that will encounter rough seas: "This is how you are to build it: The ark is to be three hundred cubits long, fifty cubits wide and thirty cubits high."[254] Modern shipbuilders use these same, or similar, dimensions when constructing barges that will likely sail through rough waters.

Although it is unclear why God chose circumcision as the sign of his covenant with Abraham (Genesis 17:11), it is of interest that circumcision has definite medical value. It also turns out that the best time to circumcise a newborn is on the eighth day after birth, when the levels of vitamin K and prothrombin (blood-clotting) are at the ideal level. Clearly it was no accident that God, the Creator of all biological systems, chose this specific day: "For the generations to come every male among you who is eight days old must be circumcised, including those born in your household or bought with money from a foreigner – those who are not your offspring."[255]

Finally, although germ theory and sterilization were not well-understood until the time of Joseph Lister, near the end of the Civil War, the Bible had already prescribed the correct procedures for dealing with childbirth (Leviticus 12), infectious diseases (Leviticus 13), bodily discharges (Leviticus 15), and the handling of the dead (Numbers 19). Those who took these commands of Scripture seriously did far better than those who neglected it.

Concluding Thoughts

The case against evolutionism is more extensive than the evidence provided in this chapter. Yet so many people today accept either an eternal universe or a chance, random origin for the cosmos, as well as a naturalistic origin of life from non-life and some type of 'upward' evolutionary scheme for all plants and animals. As Norman Geisler and Frank Turek once wrote in the naming of their book, "I don't have enough faith to be an atheist."[256] That is a sentiment I share wholeheartedly.

As I mentioned at the beginning of this book, it has been my experience that

---

[253] Leviticus 25:4-5, NIV.

[254] Genesis 6:15, NIV.

[255] Genesis 17:12, NIV.

[256] Norman L. Geisler & Frank Turek, *I Don't Have Enough Faith to Be an Atheist* (Wheaton, IL: Crossway Books, 2004).

evolutionism is not the key reason for most people rejecting the God of the Bible. Instead, it is the claim that pain, suffering, and evil proves there is no God, closely followed by the objection that Christians automatically assign to Hell those who never even had a chance to hear the message of Christ while still alive on the earth. This is the focus of chapters seven and eight. I have no doubt, however, that there are some people who do reject the life of faith primarily because they are convinced that evolution dispels the notion of God and any need for the Bible. Evolutionism is an ancient philosophy that can be used to attack both God and biblical creationism.[257] Yet some people will shout, "But maybe God used evolution as his means of creating." But in the end, this idea (theistic evolutionism) is really nothing more than an attempt to add evolutionism to the Bible, and it turns out to be an even worse idea than the atheistic version in that both biblical creationists and pure materialists equally cringe at the idea. The wedding of evolutionism and creationism can be distasteful for both sides.

But before we delve into the problem of suffering and evil as well as divine justice, let us first look at the age of the earth and when mankind was created. Although my original intention was to avoid the age debate as much as possible, this is an issue that cannot be avoided for long when examining origins. When it comes to the age of the earth and universe, many Christians are quick to jump into the fray. There seems to be no end in sight regarding this battle.

When writing the early drafts of this book, I suspect I felt like Jude, the earthly brother of Jesus. Jude's initial intention was to write a letter to the churches concerning salvation in Christ alone, but God had other plans for him: "Dear friends, although I was very eager to write to you about the salvation we share, I felt compelled to write and urge you to contend for the faith that was once for all entrusted to God's holy people."[258] Jude ended up writing a letter to urge his fellow believers to contend for the faith, by being on guard against false teachings. My intention had always been to skirt the age of the creation debate as much as possible. Instead, it took up an entire chapter – and it is the longest chapter, at that. Like Jude, I guess God had other plans for me. (At least for part of the book.)

My initial reason for wanting to avoid the age debate was quite simple: I felt like there are many creationist authors who tackle this topic better than I do. That is a bold (maybe even crazy) admission for an author, but I am being transparently honest about it. I emphasize the beginning point of the universe in my apologetic, rather than the age of the earth and universe. But then the more I thought about it, I realized something important: My beliefs about the age debate are unusual among recent creationists. I point out things that, quite frankly, no other recent creationist would dare to do. (Maybe for fear of ostracizing oneself.)

---

[257] Of course, many theistic evolutionists would never dream of using evolutionism to attack the God of the Bible. However, not every theistic evolutionist is a Bible-believing Christian, and not every evolutionist subscribes to the theistic version of evolution.

[258] Jude 3, NIV.

But the more I thought about it, I realized that maybe I should point out these different beliefs of mine.

So, how are my beliefs different from the usual recent creationist? For starters, I believe that a good case can be made for the accuracy of the Septuagint in calculating the age of humanity and (by extension) the creation. Most recent creationists cling tightly to the Masoretic numbers, but I believe that we should at least consider the Septuagint's chronology. Also, I believe that there could be some gaps in the biblical genealogies, albeit limited. (I could get rotten tomatoes thrown at me for that one!) I am convinced that old earth creationists do have a leg to stand on, even if I prefer the recent creation model for good reasons. Rather than stressing the evidence for a young creation, I am convinced that the fact of a beginning for the universe is of far greater apologetic importance. Few recent creationists seem to extol the strength of the cosmological argument for God and creation, but that line of theistic evidence is a huge part of my apologetic. I believe that the issue of animal death before the fall of Adam is a hot button topic precisely because both young earth and old earth creationists make great points – and we need to listen to each other more intently, and more sincerely. Finally, my preferred apologetic method is a combination of the classical approach and rational fideism. The presuppositional approach, which is the preferred apologetic method of almost every professing recent creationist on the planet, has merit. There is no doubt that presuppositional apologists make great points. At the request of some of my recent creationist friends, I have earnestly tried to make this method my 'go-to' apologetic – but to no avail. For me, classical apologetics combined with rational fideism works the best. I seem to be hard-wired for this approach. This was my approach to proclaiming the faith when writing both my dissertation as well as my first book, *Worldviews in Collision*. For me, it just feels natural. (I will discuss this in more depth in chapter nine.)

I trust you will find chapter six to be intriguing, whether you agree with my ideas or not. If you read with both an open mind and an open heart, you may find yourself being challenged in a healthy – and beneficial – way.

CHAPTER SIX

The Stumbling Block: A Recent Creation

In this chapter we will examine the debate over the age of the creation. Some skeptics are open to theistic evolutionism, and a few might even consider the beliefs of old earth creationists, but recent creationism is simply out of the question for skeptics. A young earth and universe are a major stumbling block for many unbelievers, as well as for some believers. As a former skeptic, I get that. We have been so conditioned to accept the belief in an old earth and universe that the idea of a creation only thousands of years old is incredulous. I found the idea to be crazy the first time I heard about it, but it intrigued me enough to investigate the issue further, to see if it had any merit. Although skeptics claim otherwise, we will see that recent creationists have good answers to the tough questions of dating the creation.

Could Creation Be Young?

I once gave a talk on creation in my home church that was a little different from my other talks on the subject. Before the talk began, I decided that I would avoid the issue of the age of the earth for as long as I could. I would only discuss it if someone in attendance first brought it up. My question to myself was, "How long will that take?" The answer? Less than five minutes. Now, many years and several talks later, I wonder what took so long. This seems to be such a hot button topic for Christians.

Here is the problem, at least as I see it: A straightforward reading of Scripture reveals a young creation. However, science in the past several decades seems to point us in the direction of an earth and universe that are billions of years old. What should we make of this discrepancy?

We must keep in mind that the age of the earth and universe has been determined by man-made methods. Therefore, they are not infallible. That makes me sound like someone who is anti-science, but rest assured that is not the case. Far from it. I do, however, believe it is not only possible but quite likely that we could be missing some things that could greatly alter how we scientifically date the creation.

Dating the creation will always involve assumptions. In the spirit of being 'fair and balanced,' both evolutionists and creationists assume things. For example, old earth creationists assume that science has proven beyond a shadow of a doubt that the earth and universe are a certain age. For them, it is 'case closed.' On the flip side, young earth creationists assume that Scripture has proven beyond a shadow of a doubt that creation is an entirely different age. For them, it is also 'case closed.' Both groups claim Scripture as their ultimate authority, and both groups claim to be fully committed to the scientific method.

Distant Starlight in a Young Universe

I have heard skeptics say that distant starlight proves the universe is billions of years old. This is, in their view, a non-negotiable. I am not convinced, however. There have been some interesting young universe models put forward in recent times: "Interestingly, biblical creationists have known about the distant starlight problem for a while and have been working on solutions."[259] I have no doubt that some of these solutions have merit. One of the most interesting solutions today comes from Russell Humphreys, a physicist who is known for his white hole model of creation. Dr. Humphreys has proposed that while time passed relatively slowly on the earth during the creation week, time flew by at highly accelerated rates elsewhere in the universe.

Helmut Welke, the founding president of the Quad City Creation Science Association, explains his preference for Dr. Humphrey's white hole model:

> I prefer Dr. Humphrey's solution because it is based on a very well-qualified theory in physics: Albert Einstein's theory of relativity. Dr. Humphrey's model of the universe – white hole cosmology – is developed with this theory and continues to be refined for those who prefer the technical details. But generally, it consists of the idea that the speed of light (and time) is much faster in deep space (away from the gravitational effect of the sun) relative to earth. For me, the simplest answer is that light does go much, much faster in deep space, away from the gravitational effect of the sun and other stars. We have measured the speed of light in a vacuum, but that is a far cry from going out beyond the orbit of Pluto and measuring it there. We already know that gravity affects time. Perhaps the two are related: time and the speed of light could be opposite sides of the very same coin, which would really bring us back to something along the lines of Dr. Humphrey's work.[260]

It is interesting to compare Dr. Humphreys white hole model[261] with Gerald Schroeder's 'two clocks' model.[262] The former holds to a young universe, while the latter generally holds to an old universe, yet they both include gravitational time dilation into their models.

There are other solutions for the distant starlight problem as well. I find Danny Faulkner's *Dasha* solution to be fascinating. *Dasha* is the Hebrew word for 'sprout.' Dr. Faulkner proposes that many creation events during the six days of

---

[259] Danny Faulkner & Bodie Hodge, "What About Distant Starlight Models?" https://answersingenesis.org/astronomy/starlight/what-about-distant-starlight-models/
[260] Email exchange between Helmut Welke and myself, November 19, 2020.
[261] Ibid.
[262] Schroeder.

Genesis took place at highly accelerated rates, which would be unthinkable today. Although lengthy, it is worth quoting Dr. Faulkner and fellow creationist Bodie Hodge on the *Dasha* solution:

> While some things were created *ex nihilo* (out of nothing) during creation week (Genesis 1:1), many things during that week probably were made of material created earlier in the week. [This is in line with Augustine's view that God created all the matter at the initial moment of creation, on day one, and then fashioned that matter into everything that exists over the next five-six days.] For instance, the day three account tells us something about how God made plants (Genesis 1:11–12). The words used there suggest that the plants shot up out of the ground very quickly, sort of like a time-lapse movie. That is, there may have been normal growth accomplished abnormally quickly. The result was that plants bore fruit that the animals required for food two to three days later. The plants had to mature rapidly to fulfill their function.
>
> God made stars on day four, but to fulfill their functions the stars had to be visible by day six when Adam was on the scene. As the normal process of plant development may have been sped up on day three, the normal travel of starlight may have been sped up on day four. If so, this rapid thrusting of light toward earth could be likened to the stretching of the heavens already mentioned.
>
> Some people may want to equate this stretching of starlight with some physical mechanism such as cdk [speed of light decay] or relativistic time effects, but this would not explain the abnormally fast development of plants on day three. This also overlooks the fact that much about the creation week was miraculous, hence untestable today. If one were to attempt to explain the light-travel-time problem in terms of a physical mechanism, one might as well look for a physical mechanism for the virgin birth or Resurrection.[263]

I could not agree more. We must remember that the creation week was miraculous in nature, not a naturalistic process. I have heard it said many times by old earth creationists that the universe could not be young because it appears to be so old. But what would an old universe look like? Old earth creationists would answer that question by invoking the issue of distant starlight: Since we can see stars billions of light years away, the universe must be billions of years old. That does make sense, but do we really know all the factors involved? Could God have stretched interstellar light at an abnormally fast rate on Day Four?

---

[263] Faulkner & Hodge.

Before addressing that question, let us examine the 'appearance of age' issue in more detail.

Skeptics of recent creationism say that if God created with the appearance of age, that makes him out to be a deceiver. I really disagree with this notion. Big time. If God created the universe – and he did – there had to be some appearance of age throughout creation, probably in every single aspect, from the very beginning. That does not make God a deceiver, it makes him an exquisite Craftsman who fashions things the way they were ultimately meant to appear, right from the beginning. If creation was a process that operated through the laws of nature, as theistic evolutionists and old earth creationists claim, then the appearance of age argument is deceptive. But if creation is a miraculous event that took place prior to the laws of nature being set in motion (as we know them now), then God is not deceptive for creating things that appeared older than they were. This idea that recent creationists are proclaiming a deceptive Creator is nonsense, and it needs to be abandoned by everyone who proclaims God's omnipotence and omniscience.

When Jesus turned water into wine (John 2:1-11), did the wine appear aged? I fully expect it did. I assume it looked like ordinary, aged wine. Yet it was only minutes if not seconds old when the guests and workers examined it with their own eyes. Was this deceptive? Hardly. For that matter, what about Adam and Eve? Did they spend the first years of their lives as babies, then toddlers and infants, before entering the turbulent teenage years? I have no doubt they looked fully mature, as adults, from the beginning of their creation.

The old earth creationist David Snoke believes that a young-appearing earth would be a divine deception: "Scientifically I would feel it utterly dishonest to argue that the world actually looks young. I would hope that those who disagree with my Bible interpretation at least agree to this; otherwise they stand open to the charge of intellectual dishonesty."[264] In my lifetime, I have been dishonest a time or two – it is all part of living with a fallen nature, in a fallen world – but I have never been intellectually dishonest about thinking the earth looks a certain age. My point has always been this: There had to be an appearance of age in a brand-new creation that was fully mature in every way.

We all need to be aware of the difference between functional maturity and age. Something can be functionally mature but chronologically young at the same time. In the movie *Genesis: Paradise Lost*, during the creation week trees are shown to have 'sprouted' from the earth so rapidly that it makes the usual time-lapse photography look slow by comparison. (This is an example of the *Dasha* solution.) Within a moment of their completed creation, these trees appear as though they were decades if not hundreds of years old, yet they are only seconds old. Likewise, God could have stretched the light between solar bodies so quickly that any chance of using modern methods to measure the distance in light-years between these bodies is totally inaccurate. Modern methods of

---

[264] David Snoke, *A Biblical Case for an Old Earth* (Grand Rapids, MI: Baker Books, 2006), 48.

measuring the distance between astronomical bodies only works when several assumptions are correct.[265] However, since we do not know exactly how God created – that is, in the details of creation that are beyond the general outline given in Genesis 1 – it is safe to say that the universe can be both extremely young and incredibly massive at the same time.

This is going to be a bold statement, especially for a non-cosmologist, but cosmology seems to be heavy on philosophy – although clearly a lot of science has been incorporated as well. Vast distances between stars could point to a universe billions of years old, but keep in mind that the apparent age of the universe does not necessarily equate to a real or true age.

If the creation began from an infinitesimal 'point,' if the speed of light has remained constant from the very beginning of the universe to the present, and if the rate of the universe's expansion has always been the same as well, then I agree that the creation is billions of years old. But those are three big 'ifs.' There is much that we do not know, and will likely never know, about the initial conditions of the universe. What if the universe began not as an infinitesimal point, but as a much larger structure? The expansion of the universe would have begun from a different starting point, and therefore it would be much younger than secular cosmologists say it is. And what if the speed of light was considerably faster during the creation week? Again, this would make the universe younger than secular cosmologists say it is. And what if God 'stretched the heavens' at an expanded rate during the creation week compared to what is observed by astrophysicists today? Again, this would make the universe younger than secular cosmologists say it is. And what if two or even all three of these secular assumptions are wrong? If that is the case, the universe could be quite young. I have no problem with a young universe, and neither should anyone else: There are so many 'ifs' in cosmology that it cannot be ruled out. A recent creation does go against everything we were taught by secular academia and the media, but it should not be quickly dismissed as it so often is. Recent creationism is a legitimate interpretation of Scripture, and despite what skeptics say the scientific data does not necessarily rule it out.[266]

John Eddy (1931-2009) was an American astronomer who specialized in solar activity. In 1978, Dr. Eddy made a statement that caught the attention of not only astronomers but creationists worldwide:

---

[265] If we knew for certain that the universe began from an infinitesimal point, perhaps the size of the period at the end of this sentence, and the speed of light has been constant from the beginning of time, then I agree that the universe is billions of years old. However, those are two significant assumptions that cannot be known with absolute certainty.

[266] It should be noted that secular Big Bang cosmologists also have a light-travel time problem, which is commonly referred to as the 'horizon problem.' See Jason Lisle's "Light-Travel Time: A Problem for the Big Bang," which may be found at Creation Ministries International (creation.com).

> I suspect…that the Sun is 4.5 billion years old. However, given some new and unexpected results to the contrary, and some time for frantic readjustment, I suspect that we could live with Bishop Ussher's value for the age of the Earth and Sun (approximately 6,000 years). I don't think we have much in the way of observational evidence in astronomy to conflict with that.[267]

This is exactly how I feel! Let me say it again: I believe that Scripture points us in the direction of a recent creation, and I also believe that the science involved does not necessarily prevent that view. There are scientific indicators for an old creation, but there is also scientific data that could lead us to believe that the creation is much younger than we have been told. There are so many factors involved, and some factors that we may not have even thought about yet, that like Dr. Eddy I believe it is really anyone's game.[268] I am going with my view of Scripture on the age of the creation. If you believe the earth and universe are billions of years old, that is fine with me. I hope we can still hang out.

George Francis Rayner Ellis is a high-profile cosmologist who worked with the late Stephen Hawking, who is considered by many to have been the greatest cosmologist in modern times. Dr. Ellis admitted that philosophical assumptions are a big part of cosmology:

> People need to be aware that there is a range of models that could explain the observations…For instance, I can construct you a spherically symmetrical universe with earth at its center, and you cannot disprove it based on observations…You can only exclude it on philosophical grounds.[269]

When it comes to dating the creation, it really is anyone's game. By 'range of models' we can reasonably include Dr. Schroeder's 'two clocks' model, or Dr. Humphrey's white hole model, or any recent creationist model. Of course, this range of models would also include anything devised by Hugh Ross or any other old earth creationist – we must be fair, after all. Dr. Ellis is saying, at least as I understand him, that when it comes to cosmology, we can never be too sure about our cherished model. Therefore, I am going with Scripture. It is better to trust the One who was there in the beginning and told us about it.

I highly recommend Spike Psarris' three-volume DVD set *What You Aren't Being Told About Astronomy*. These videos are a must-have for anyone interested

---

[267] John Eddy. Geotimes Magazine (September 1978), 18. Report on Symposium at Louisiana State University.

[268] Along this line, we must ask the question, "Did God create everything through the laws of nature, or did God create the universe and then set the laws of nature into motion?" These two options alone have huge implications regarding the age of the earth and universe.

[269] W. Wayt Gibbs, "Profile: George F.R. Ellis: Thinking Globally, Acting Universally." Scientific American, Volume 273: Number 4 (October 1995).

in discovering the truth about cosmology.

Radiometric Dating: A Question of Trustworthiness

Cosmology as indisputable proof for an old universe does not work for me. Likewise, radiometric dating as indisputable proof of an old earth does not work for me, either. Radiometric dating is based upon several starting assumptions that are just that: Assumptions. Mike Riddle, one of the most impressive creationist speakers today, notes the inaccuracy of radiometric dating:

> A rock sample from the newly formed 1986 lava dome from Mount Saint Helens was dated using Potassium-Argon dating. The newly formed rock gave ages for the different minerals in it of between 0.5 and 2.8 million years. These dates show that significant argon (daughter element) was present when the rock solidified.[270]

This is just one example demonstrating erroneous age estimates that may be obtained by radiometric dating. Many other examples could be included, but the point is clear: Radiometric dating may not be the exact science that people have been led to believe it is.

The geologist Andrew Snelling notes three primary assumptions behind radiometric dating. First, we cannot know with certainty the conditions present at 'time zero':

> No geologists were present when most rocks formed, so they cannot test whether the original rocks already contained daughter isotopes alongside their parent radioisotopes. For example, with regard to the volcanic lavas that erupted, flowed, and cooled to form rocks in the unobserved past, evolutionary geologists simply assume that none of the daughter argon-40 atoms was in the lava rocks…yet lava flows that have occurred in the present have been tested soon after they erupted, and they invariably contained much more argon-40 than expected.[271]

This single assumption can throw the dates off by orders of magnitude. This explains why rocks of a known age, that were formed in the very recent past, can date to millions of years old.

Second, it is generally assumed that the rocks being dated have no

---

[270] Mike Riddle, "Does Radiometric Dating Prove the Earth Is Old?" https://answersingenesis.org/geology/radiometric-dating/does-radiometric-dating-prove-the-earth-is-old/

[271] Andrew Snelling, "Radiometric Dating: Problems with the Assumptions." https://answersingenesis.org/geology/radiometric-dating/radiometric-dating-problems-with-the-assumptions/

contamination. No one can really know if this is the case, however:

> The problems with contamination, as with inheritance, are already well-documented in the textbooks on radioactive dating of rocks. Unlike the hourglass, where its two bowls are sealed, the radioactive 'clock' in rocks is open to contamination by gain or loss of parent or daughter isotopes because of waters flowing in the ground from rainfall and from the molten rocks beneath volcanoes. Similarly, as molten lava rises through a conduit from deep inside the earth to be erupted through a volcano, pieces of the conduit wall rocks and their isotopes can mix into the lava and contaminate it…Because of such contamination, the less than 50-year-old lava flows at Mount Ngauruhoe, New Zealand yield a rubidium-strontium 'age' of 133 million years, a samarium-neodymium 'age' of 197 million years, and a uranium-lead 'age' of 3.908 billion years![272]

Third, a constant decay rate is assumed in radiometric dating. Although I have quoted Dr. Snelling's article in large part already, this section is worth quoting in its entirety so as not to miss the point:

> Physicists have carefully measured the radioactive decay rates of parent radioisotopes in laboratories over the last 100 or so years and have found them to be essentially constant (within the measurement error margins). Furthermore, they have not been able to significantly change these decay rates by heat, pressure, or electrical and magnetic fields. So, geologists have assumed these radioactive decay rates have been constant for billions of years.
>
> However, this is an enormous extrapolation of seven orders of magnitude back through immense spans of unobserved time without any concrete proof that such an extrapolation is credible. Nevertheless, geologists insist the radioactive decay rates have always been constant, because it makes these radioactive clocks 'work'!
>
> New evidence, however, has recently been discovered that can only be explained by the radioactive decay rates not having been constant in the past. For example, the radioactive decay of uranium in tiny crystals in a New Mexico granite yields a uranium-lead 'age' of 1.5 billion years. Yet the same uranium decay also produced abundant helium, but only 6,000 years worth of that helium was found to have leaked out of the tiny crystals.

---

[272] Ibid.

This means that the uranium must have decayed very rapidly over the same 6,000 years that the helium was leaking. The rate of uranium decay must have been at least 250,000 times faster than today's measured rate! For more details see Don DeYoung's *Thousands...Not Billions* (Master Books, Green Forest, Arkansas, 2005), 65-78.[273]

Nonetheless, every old earth creationist book or paper I have ever read has been adamant that radiometric dating is reliable. Roger Wiens is one of those old earth creationists. Dr. Wiens holds a doctorate in physics, along with a minor in geology. His doctoral thesis was on isotope ratios in meteorites, including surface exposure dating. He has worked with Caltech's Division of Geological & Planetary Sciences as well as the Space & Atmospheric Sciences Group at the Los Alamos National Laboratory. In other words, he knows his stuff. Dr. Wiens offers his view on the accuracy of radiometric dating:

> There are well over forty different radiometric dating methods, and scores of other methods such as tree rings and ice cores. All of the different dating methods agree – they agree a great majority of the time over millions of years of time. Some Christians make it sound like there is a lot of disagreement, but this is not the case. The disagreement in values needed to support the position of young-Earth proponents would require differences in age measured by orders of magnitude (e.g., factors of 10,000, 100,000, a million, or more). The differences actually found in the scientific literature are usually close to the margin of error, usually a few percent, not orders of magnitude![274]

Ralph Muncaster, an engineer who has authored several books on apologetics, agrees with Dr. Wiens: "The accuracy of radiometric dating is very high. The half-lives of most radioactive isotopes used for dating are known to within plus-or-minus 2 percent."[275] I used to work with a physicist who holds to the old earth view. Like everyone in his camp, he was convinced that the accuracy of radiometric dating is beyond question. Despite this objection from my old earth brethren, the assumptions involved in radiometric dating are very real. Therefore, I am not convinced by radiometric dating as indisputable proof for an old earth.

As a radiation therapy student in the late eighties, I was frequently given the task of measuring the radioactive background of used medical radioisotopes. One

---

[273] Ibid.
[274] Roger Wiens, "Radiometric Dating: A Christian Perspective."
   https://d4bge0zxg5qba.cloudfront.net/files/articles/non-staff-
      papers/roger_wiens_radiometric_dating.pdf/
[275] Ralph Muncaster, *Dismantling Evolution: Building the Case for Intelligent Design* (Eugene, OR: Harvest House Publishers, 2003), 232.

day, my supervising physicist told me, before I had done anything, what the results would be. My first thought was, "Seriously?" My second thought was, "Why even bother doing this?" (But, since I was the student and he was the boss, I did it.) Interestingly, the results were not what he said they would be. I did the math three times, but I was consistently off from what I was told I would find. I reported my results to him, and his solution was simple: "Your math is off!' He checked it, and it turned out that I had been correct. He was dumbfounded. That was a valuable lesson that has served me well ever since: Scientists assume things, just like all of us do. Most scientists want to know the truth about the natural world, but nonetheless they may interpret the scientific data through the lens of their worldview – even if inadvertently.

Death Before Sin

Fossils are another supposed 'proof' for an old earth, but upon closer inspection fossils demonstrate creationism. Transitional fossils, or 'missing links,' are those fossils or fossil series that supposedly connect one distinct type of plant or animal to another. Unfortunately for evolutionists, the missing links are still that: Missing. A lack of clear-cut transitional forms in the fossil record indicates that the various life forms were created individually (Genesis 1:11-12, 20-21, 24-27). Not only are there no evolutionary ancestors in the beginning of the fossil record, as seen in the Cambrian explosion, but also no transitional forms appear after the initial appearance of the major body shapes. Concerning animals, all the supposed transitional forms can be refuted, such as archaeopteryx (supposed reptile-bird transition) and ambulocetus (supposed land mammal-whale transition). But for now, let us skip over the scientific concerns associated with fossils, and instead dive into the theological consideration of the 'record of death.'

Evolutionists claim that the earth's fossil layers date back millions of years. But if one allows for long ages, is he or she has accepting death, disease, and suffering before the fall of man? Recent creationists reject both human and animal death before the fall of Adam, while old earth creationists accept only animal death before the fall; neither camp is comfortable with human death before the fall. On the other hand, theistic evolutionists maintain that death has always been part of the creation, including the death of hominids who may have been anatomically similar – if not identical – to Adam. Death before the fall is a huge debate between recent creationists, old earth creationists, and theistic evolutionists. Is death a consequence of man's sin, or did death exist before Adam and Eve fell? If so, did that include human death? Let us look at the recent creation view first.

If the Garden of Eden were sitting on fossil layers millions if not billions of years old, then blood was shed long before human sin entered the world. For recent creationists, this belief destroys the foundation for Christ's sacrificial atonement, since the Bible clearly teaches that the sin of Adam brought death and suffering into the world. Romans 8:19-22 tells us that the whole of creation 'groans' because of the effects of the fall of mankind. It is not just mankind who 'groans' or suffers, but everything within the creation is in a state of despair because of sin. This includes the animals, of course. But if animals were ripping each other apart and chowing down on the remains of their victims before Adam and Eve were on the scene, then suffering and death existed in the world before the fall of man. That would make death a normal part of the creation, rather than death being an 'enemy' (1 Corinthians 15:26).

At the conclusion of the six-day creation week, God pronounced his work as 'very good' (Genesis 1:31), since the fall of mankind had not yet taken place. It seems strange to me that God would call his creation 'very good' if death, disease, and suffering had already entered the world by the time of Adam and Eve. Paul, writing under the inspiration of the Holy Spirit, revealed that, "sin came into the world through one man, and death through sin, and so death spread to all men because all sinned."[276] Sin did not enter the world before Adam's rebellion, and death came through sin, yet evolutionism requires death on an unimaginable scale long before mankind's arrival, to bring about the various plants and animals over eons of time. (The death, or extinction, of some creatures paves the way for the rise of other creatures, according to evolutionism.) Were Adam and Eve enjoying God's 'very good' creation in the Garden of Eden, despite standing upon layers of dead and buried creatures? God's pronouncement of a 'very good' creation and the Darwinian notion of nature being 'red in tooth and claw' is difficult for me to reconcile.

Fortunately, the creation will be liberated from its bondage to decay, eventually being brought into a state of freedom and glory. Revelation 21-22 reveals that there will be a 'new heavens and a new earth' in the future, where there will be no more death – just like it was before sin entered the world. Claiming millions of years for the earth and universe presents a theological difficulty for the foundations of the Gospel message.

Old earth creationists beg to differ, of course. For them, animal death existed before Adam and Eve fell, but mankind did not experience death before the fall. An email exchange between myself and a good friend of mine – my primary mentor and retired pastor, Lee Johnson – highlights the old earth position on death before the fall of man:

> I suppose one thing I think you may have jumped too far on is your belief that the 'death' God warned Adam about included animal death. It just seems inconceivable to me that life forms created by God to multiply, in many cases exponentially, could have sustained life on earth without some mechanism in place to remove the excess. Animals do appear to have been created to eat, and to be eaten by, other animals. What warning would it have

---

[276] Romans 5:12, ESV.

been to Adam about him dying if there was not already death? Is animal death a bad thing inherently or just part of the process God built into the system?[277]

These are good points, and recent creationists need to be ready to address them in the spirit of 'gentleness and respect' (1 Peter 3:15). Is death the only way that an excessive population of animals could be dealt with? Likewise, if man had not fallen – and did not experience physical death as we do now – it seems reasonable to assume that people eventually would have experienced a population problem. God commanded the first people to, "Be fruitful and increase in number; fill the earth and subdue it."[278] But the earth is only so big, and if people did not experience physical death the planet would have been overloaded in a short time. How might that situation have been dealt with? Perhaps in an unfallen world, people would eventually be taken up to Heaven directly, as Enoch was (Genesis 5:24). That is a strong possibility. If so, could the same situation have happened to animals? I can hear it now: "Nonsense!" There is a case to be made for animals in the afterlife, however. Although it is beyond the scope of this study, I will point you in the direction of a fantastic resource in this area: Dan Story's *Will Dogs Chase Cats in Heaven? People, Pets, and Wild Animals in the Afterlife*. Or, as my old earth friends insist, why not just accept animal death before Adam as part of reality? Maybe God really does 'have teeth,' so to speak. But maybe recent creationists are correct, and animals did not have a chance to die in the short time before mankind fell.

Theistic evolutionists have no problem with animal death, or even the death of beings closely resembling Adam, prior to the fall of man. I can see the points that old earth creationists make regarding this issue, although as I already mentioned I struggle with their view, but I thoroughly oppose the theistic evolutionary viewpoint. The idea of hominids – that is, ape-men – prior to Adam is unscriptural and scientifically unconvincing. I will listen to the beliefs of old earth creationists, but theistic evolutionism simply does not work for me. Although I will respect the theistic evolutionist's civic and God-given right to believe as they see fit, I disagree with their understanding of creation. I know people who are theistic evolutionists, and they are wonderful people and good theologians in other ways, but evolutionism is evolutionism – whether God is in the picture or not. Evolutionism simply does not make sense, theologically or scientifically.

Old earth creationists make another good point: Animals appear designed to bite, claw, and generally rip each other to pieces. Many animals seem to be perfectly designed killing machines. Recent creationists suggest, however, that God – in his perfect foresight – designed animals with these ferocious characteristics because he knew they would be needed in a fallen world that is

---

[277] Email exchange between Lee Johnson and myself, October 14, 2020.
[278] Genesis 1:28, NIV.

'red in tooth and claw.' This brings up a host of other theological questions, such as why did God not prevent the fall in the first place? Here are the possibilities: One, God designed animals with the 'equipment' and the instinct to fight and kill because animal death was always intended as a way of life in this world, even before the fall of man. This is the old earth creation view. Two, God designed animals with the equipment needed to fight and kill from the beginning of their creation because God knew that they would need these abilities to survive in a fallen world, which he knew was coming. This is a popular recent creationist view. Three, God designed animals with the equipment needed to fight and kill in case man fell and compromised the creation. This is another recent creation view. (In the second view, God knew the fall was coming, while in the third view God chose not to know the future, therefore he designed animals with the necessary survival abilities in case they were needed in a fallen world.) Four, after the fall of man God adapted, or re-designed, many animals to survive in this harsh world. In other words, God re-designed them for life in a now-fallen world. This is another recent creationist view, although I am far from being sold on the idea. Five, animals evolved these characteristics out of necessity. The animals that evolved the ability to fight and kill effectively survived this harsh world. This is the theistic evolution view, which I do not agree with.

Helmut Welke explains a sixth view:

> My own view is that the teeth and claws were designed for peaceful living in a very flourishing world. The best single example is the Panda. This calm animal has very sharp claws and is an expert tree climber. It has very sharp canine teeth, but its diet is 99% bamboo. They may occasionally eat some fruits. Animals with sharp claws can get traction and run fast (like cheetahs), as well as climb trees. Sharp claws and teeth were designed to rip away at tough vegetables and fruits, as well as bamboo. The teeth of a T-Rex, (which scientists now consider to be a slow-moving scavenger and not a predator) could use those teeth to munch on watermelon, cantaloupes and tough foods like avocados and gourds. Most likely they could also chew on certain types of tree bark. Sharp canines are necessary for that as well. But sin entered the world, and everything went downhill. People and animals learned to prey on one another. Like any tools, these claws and teeth can be used for both good and bad. This was accelerated after the Flood when the food supply was different and not as plentiful. Note that it was after the Flood that God instructed Noah that they could now eat meat as a food source. This, in a nutshell, is my favorite explanation.[279]

---

[279] Email exchange between Helmut Welke and myself, November 19, 2020.

That is a great explanation which dovetails perfectly with God's 'very good' design. In this view, animal design is not so concerned with the fall, as it is with practical, day-to-day living.

Table 6.1

Views on Animal Death Before the Fall

VIEW #1: OEC

DEATH WAS ALWAYS PART OF GOD'S CREATION.

VIEW #2: YEC

GOD KNEW THE FALL WAS COMING; GOD WAS READY.

VIEW #3: YEC

GOD WAS NOT SURE IF MAN WOULD FALL; GOD 'COVERED THE BASES.'

VIEW #4: YEC

GOD REDESIGNED ANIMALS FOR A FALLEN WORLD.

VIEW #5: TE

ANIMALS EVOLVED FIGHTING ABILITIES OUT OF NECESSITY.

VIEW #6: YEC

TEETH, BEAKS, AND CLAWS WERE

ORIGINALLY MEANT FOR NON-HARMFUL PURPOSES.

Is it possible that animal death occurred before man fell, but in a young creation? We have no idea how long there was between Adam and Eve's creation and their fall. Maybe it happened quickly, just a short time after they were created, or maybe it took years. No one knows, save God. In the time between creation and the fall, is it possible that at least one animal died by accident? I am not talking about being ripped to shreds by a predator, but merely an accident, such as falling off a cliff, drowning, or so forth. Is it possible that a large land-dwelling animal stepped on a small creature by accident? If none of these situations happened before the fall of man, that means God oversaw every aspect of every creature's life, preventing accidents on a continual basis. This is not out of the question in a world created by the sovereign God, of course, but is it likely? Some recent creationists argue that the pre-fallen world was dramatically different from the world we now live in, but did that include God overturning natural law at nearly every moment? Accidents happen a lot today, but were they divinely prevented in the pre-fallen world? Food for thought. (Let me leave it at that.)

When God told the recently fallen Adam that he would eventually 'return to the ground' or die (Genesis 3:19), did Adam really comprehend death? If so, was it merely an intellectual understanding only, or was it experiential as well? If you are a recent creationist, I want you to think deeply about these things. I am not saying that animal death before Adam's fall happened, but I am saying that old earth creationists make some valid points in this area. Never fear examining both sides in depth.

Animal death before Adam is not a side issue in creationism: This is a huge area of debate, which calls into question both God's character and the reliability of Scripture. Creationists from both camps have engaged in some serious battles over this topic. The recent creationist view makes theological sense to me, but I totally get where my old earth friends are coming from, which is why I point out their objections to the recent creation view. I recommend researching this topic in more detail. I have only given you a brief introduction to an important issue. I urge you to consult both young earth and old earth resources on the topic, and I hope you read with an open mind and go where the evidence leads.[280]

But Only 6,000 Years Old?

Although a young creation makes theological sense to me, I have always struggled with one thing: The amount of time between the Flood and the Babel event seems too brief to accomplish all that needed to take place. Between the Flood and the birth of Peleg, there is only 101 years according to the genealogies given in the Masoretic text (assuming no gaps). Let us examine the genealogy from Adam to Abraham, to understand my point.

Keep in mind that the following genealogy is based upon the Masoretic text. This is an important point that we will come back to later. The list uses the *Anno Mundi* calendar, which is Latin for 'in the year of the world.' The year given is based upon Adam being created in the year zero, or at the creation itself. (For old earth creationists, this is assuming a lot.) Seth, the son of Adam and Eve, was born when Adam was 130 years old. Therefore, Seth's birthdate is *Anno Mundi* 130, or simply AM 130. The genealogy is as follows:

Adam, 0; Seth, AM 130; Enosh, AM 235; Kenan, AM 325; Mahalalel, AM 395; Jared, AM 460; Enoch, AM 622; Methuselah, AM 687; Lamech, AM 874; Noah, AM 1056; Shem, AM 1559; Flood, AM 1656-1657; Arphaxad, AM 1659; Shelah, AM 1694; Eber, AM 1724; Peleg, AM 1758; Tower of Babel, AM 1758 (Or Shortly Thereafter); Reu, AM 1788; Serug, AM 1820; Nahor, AM 1850; Terah, AM 1879; Abram, AM 2009.

From the Flood to Babel, it always seemed to me that there was little time for a significant population to have arisen. Keep in mind that the Tower of Babel was quite an achievement, and it would have required many able-bodied builders. After all, it was impressive enough to warrant God's intervention. The Flood ended in AM 1657, and we know that the event at Babel – in which the people scattered upon the face of the earth from there, according to their languages (Genesis 11:1-9) – happened just a short time later, during the time of

---

[280] For resources on the young earth position, I recommend searchcreation.org, and for resources on the old earth position I recommend both reasons.org and godandscience.org. In each case, simply type 'animal death before Adam' into the search bar.

Peleg. Moses wrote that, "Two sons were born to Eber: One was named Peleg, because in his time the earth was divided; his brother was named Joktan."[281] If we base the gap between these two events on the beginning of Peleg's life, there was a mere 101 years from the Flood to Babel. I always assumed that we should base the gap on the beginning of Peleg's life, since Peleg's father Eber named his son in recognition of this division. (Eber, also known as Heber, would have been present at Babel. His language group is believed to have been named after him: Heber, or Hebrew.) On the other hand, if we base the gap between the Flood and Babel on the end of Peleg's life, there was 340 years between these two events. That number seems more reasonable to me. Really, the gap between the Flood and Babel could be any number between 101-340 years, although as mentioned I always assumed it was the lower number. My question was always, "Was there enough time to accomplish all that the Bible says happened during that time?" (Keep in mind that, at this point in the discussion, I am not considering the possibility of gaps in the genealogical record. We will get to that in a moment.)

After the Flood, which was devastating – every recent creationist on the planet will tell you the same thing – the world needed to recuperate and rebuild itself, so to speak.[282] Vegetation would have been totally denuded in some areas, perhaps even over large portions of the globe. What vegetation remained intact required time to spread, to turn the earth green again. (Keep in mind that the animals are eating this vegetation the whole time, which is a significant factor as well. Did that slow the spread of animal populations after the Flood, at least for a time?) I have no doubt that Noah carried plants on board the ark, and thereby contributed to the spread of vegetation as well, but even that would have taken time. The earliest people groups, which sprang forth from the lineage of Noah, needed time to develop in numbers so they could build the city and tower of Babel. Then, after the scattering from Babel, the various groups needed time to make their way around the globe, founding the various civilizations that we know about today and developing those cultures to a high level by the time of Abraham. Abraham, from Ur of the Chaldees (Genesis 11:31), was no backwoods farmer. The Chaldeans were an advanced culture, known for their expertise in astrology.[283] They were not underdeveloped cave dwellers who barely survived day-to-day living. Yet, according to a straightforward interpretation of the Masoretic text, from the Flood to Babel was a very brief time. Likewise, from the birth of Peleg to the birth of Abraham was only 251 years. This is better than the 101 years from the Flood to the birth of Peleg, but still not a lot of time to accomplish all that was necessary for human history. I always had trouble

---

[281] Genesis 10:25, NIV.

[282] It is possible, of course, that God could have miraculously healed the planet in the wake of the Flood.

[283] In the ancient world, astrology was a combination of occultism and astronomy. Ancient astrologers were considered scientifically adept in their time.

accepting that so much happened in the 352 years from the end of the Flood to the birth of Abraham.

I once had a fellow believer tell me, "I pray that someday you will have the faith to accept a strict 6,000-year history for the creation." This is not a faith issue for me. If a lack of faith is the problem, I could simply side with secular historians or old earth creationists and insert as much time as I want to between the Flood and Babel, and between Babel and Abraham. I have no problem with a young creation. My problem has always been what I perceived to be a lack of time during this stage of human history.

I was fortunate enough to discover two possible solutions to the problem, however. First, the Masoretic text may not be the only translation that creationists should consider using. In a Facebook page, the filmmaker Nathan Hoffman directed me to one of his online videos, *Were the Pyramids Built Before the Flood?*[284] It turns out that the Septuagint, the Greek translation of the Old Testament dated to 250 BC, provides 650 additional years right where they are needed. The Masoretic text, which was translated between the seventh and tenth centuries AD,[285] may not have been based upon the original Hebrew copy of the Old Testament that the Septuagint, the Samaritan Pentateuch, and the Jewish-Roman historian Flavius Josephus used. Between the Flood and Abraham, the Masoretic text left off a hundred years from each of six names (Arphaxad, Shelah, Eber, Peleg, Reu, Serug), and a seventh name (Nahor) was shorted 50 years (Genesis 11:10-25). According to the Septuagint, Arphaxad begot Shelah when he was 135 years old, not just 35 years old as the Masoretic text states. Add a hundred years to the next five names as well. Additionally, Nahor became the father of Terah at 79 years of age according to the Septuagint, not 29 years of age as the Masoretic text states. This additional 650 years gives a more realistic timeframe to early post-Flood history. The video is only about a half hour long, and well worth watching. I highly recommend that you watch it now, before continuing further.

In addition to these longer lifespans in the Septuagint, there is also the possibility of missing names in the genealogy lists. Recent creationists hate to consider this possibility, but nonetheless a case can be made for the 'telescoping' of names. Personally, I believe that abbreviated genealogies are limited, however. There is no way we can get more than a few thousand years from this at best, and certainly not tens or hundreds of thousands of years as some old earth creationists and theistic evolutionists believe. Consider the following information:

---

[284] Nathan Hoffman, *Were the Pyramids Built Before the Flood?*
https://www.youtube.com/watch?v=VI1yRTC6kGE&list=PL6P6ysO2XSKIcu--7aV8fB_78JrTKL_al&index=6
[285] The oldest extant copies date to the ninth century AD.

As evidence the genealogies are telescoped (compressed or abbreviated), scholars point to examples such as the genealogy of Moses, which appears four separate times in Scripture (Exodus 6:16-20, Numbers 26:57-59, 1 Chronicles 6:1-3, 23:6-13). Moses' genealogy is given as Levi to Kohath to Amran to Moses. As straightforward as this seems, related Bible passages suggest that several generations were likely skipped between Amram and Moses. 1 Chronicles 7:20-27 provides a parallel genealogy of Ephraim, son of Joseph (brother of Levi), from the same period of history as the Mosaic genealogies. While only four generations are listed from Levi to Moses, 12 generations [are] listed from Joseph to Joshua during the same time period [Joseph, Ephraim, Beriah, Raphah, Resheph, Telah, Tahan, Ladan, Ammihud, Elishama, Nun, Joshua].[286]

This comparison of lists demonstrates that abbreviated genealogies are found in Scripture. I realize that some people will be uncomfortable with this idea, but please know that it is not my intention to make anyone uncomfortable. If shortened lists are found at the time of Moses and Joshua, then it is wholly conceivable that there could be some genealogic abbreviation from an earlier period as well. John Greene, an old earth creationist, claims, "It has been suggested that the Mosaic genealogies are perhaps only 20 to 40 percent complete. Those who hold that the genealogies are telescoped place the creation of Adam and Eve around 10 to 30 thousand years ago, but perhaps as late as 60,000 years ago."[287] I have no problem with the 10,000-year figure, and neither did Henry Morris and John Whitcomb, the founders of the modern creation science movement. Robert Carter and Chris Hardy, recent creationists, note this in their article on the age of the earth:

> Note that the only way to get a 'traditional' date of creation of approximately 4000 BC is to use the Short Sojourn calculation and minimal to simple-additive adjustment parameters. This makes it likely that the earth is several hundred years older than most biblical creationists expect. However, we reject the idea that there are 'missing generations' that might increase the age to as much as 10,000 years, as Whitcomb and Morris did in their seminal and influential book *The Genesis Flood*.[288]

As much as I appreciate Dr. Carter – he is one of the top creationist speakers

---

[286] Greene.
[287] Ibid.
[288] Robert Carter & Chris Hardy, "The Biblical Minimum and Maximum Age of the Earth." https://creation.com/biblical-age-of-the-earth

in the world today – I believe that we need to be open to the possibility of some limited genealogic abbreviating. If Morris and Whitcomb were open to that possibility, then so am I.

Having said that, however, I do believe that the pre-Flood list is either complete as it is, or close. Jude, the earthly brother of Jesus, tells us that Enoch was "the seventh from Adam,"[289] so if there is a missing name it would likely be between Methuselah and Noah.[290] This seems doubtful to me, although I cannot rule it out. Likewise, archaeologists and historians are in general agreement that Abraham lived around 4,000 BC, and history from that time on is well-established. That leaves the period between the Flood and Abraham as the only questionable stretch of time for me. However, the Septuagint combined with the possibility of some limited genealogic abbreviating allows for adequate time. Even Dr. Carter, who favors the Masoretic text, states that, "the earth cannot be more than 7,680 years old."[291] For me, that figure is certainly much more reasonable than only 6,000 years.

The second solution, that was brought to my attention by my friend Helmut Welke, is based upon the conventional dating found in the Masoretic text. Welke has done his homework in this area. (Actually, he has done his homework in every area of the origins debate! I know of no other creationist who is better prepared than he is.) Welke makes the point that 'in the days' (KJV) or 'in the time' (NIV) of Peleg, as revealed in Genesis 10:25, almost certainly refers to Peleg in the prime of his influence, rather than Peleg's birth. A mid-range date for 'Peleg's days' might be 140 years after his birth. Taking into account that he was born 101 years after the Flood, this time would round to 240 years after the Flood. Welke, as an engineer with a propensity for mathematics, 'ran the numbers' concerning an early post-Flood population. He came up with an estimated population of about 500,000 people in the days of Peleg, based upon some reasonable assumptions. First, the following birthrate was assumed: 2.5 children by the time people reach 20 years old, 8 children by the time they reach 30 years old, and 12 children by the time they reach 40 years old. Additionally, Welke assumed that wars and murder were not so rampant that the population was kept low. Although we could not make this assumption concerning the pre-Flood world (Genesis 6:5), it is reasonable to assume that the early post-Flood people cooperated better than their pre-Flood ancestors (Genesis 11:1-6.) Due to their still-long lifespans and social cooperation, these assumptions are quite reasonable.

With half a million people on the planet, there would have been enough able-

---

[289] Jude 14, NIV.

[290] Some old earth creationists might counter this claim by stating that Jude was only commenting on the genealogy list that he had available to him, which is the same list that we have today. For me personally, however, this verse from Jude is a New Testament confirmation of an Old Testament truth.

[291] Carter & Hardy.

bodied people to build a city and a tower – which, by the way, was not completed. (Technically, they only had to begin the building project.) Additionally, the city would not be an expansive metropolis like modern-day New York City or Chicago: Ancient cities were much smaller than this. Therefore, a strong case can be made for an early post-Flood timeframe using the Masoretic numbers.

However, could it be possible that both solutions are simultaneously correct? Maybe the Septuagint numbers are right. Maybe there was some genealogical abbreviating (albeit limited) for the early post-Flood period. Maybe the Babel event happened near the mid-point of Peleg's life, rather than at his birth, allowing for a larger population to build Babel. (This makes sense, regardless of which text you prefer.) Maybe Welke's population numbers are spot-on. (They are computer simulated, after all.) That is a lot of maybe's, but they are all reasonable. Ultimately, I came to this conclusion concerning the early post-Flood population: The biblical account is trustworthy and reliable. You might prefer a strict adherence to 6,000 years – Welke's case for the Masoretic chronology is convincing – or maybe you find the Septuagint chronology more likely, with an age of the earth closer to 8,000 years. (Giving more time between the Flood and Babel, and between Babel and Abraham.) Either way, Scripture is reliable. Secular historians will tell you otherwise, but it is better to trust the One who was there and wrote the book about it.

It strikes me as funny that I have no problem with time before Adam, but I always struggled with what I perceived to be a lack of time between the Flood and Abraham. Most people wrestle over the question of what happened prior to Adam: Was there only five 24-hour days before Adam's creation, or was there billions of years? For me personally, I need to make sense of history from the time of Adam on. Human history must be long enough to account for all the Bible says happened from Adam and Eve to the present. Before Adam's creation, there is so much debate and confusion about what really happened that I decided to just let Scripture speak plainly to me. And what does Scripture say to me? God created everything in six days. Someday when I get to Heaven – and as a cardiac patient, that thought is never far from the forefront of my mind – I can ask God if it was six days or billions of years. (I will probably care less either way at that point.) If I find out it was billions of years, my response will be, "Okay." And if I find out it was six 24-hour days? The same response. Heaven will not depend upon my passing that theology exam.

As a recent creationist, I am sort of an anomaly. I was trained under old earth creationists, and to this day I thoughtfully consider the beliefs of my old earth brethren far more than any other recent creationist you will likely encounter. I really am in a strange position among creationists. My friends who are old earth creationists must think to themselves, "This guy has gone off the deep end. He seemed promising at one time." On the other hand, my young earth creationist friends must believe that I am far too soft on old earth creationists – the dreaded 'compromiser' that no one wants to be. But one thing I have learned from my studies in theology is this: There is an awful lot that all of us do not know for sure. It is quite possible that we can be wrong about some things! As a human being, I have serious limitations. Although I have a background in the natural sciences, I am not a professional scientist by profession. Likewise, although I have graduate degrees in theology, one thing every theologian should understand is that there is an infinite number of things we do not know about God and our relationship to him.

Augustine cautioned us against rushing into a position without first looking at all the facts:

> Usually, even a non-Christian knows something about the earth, the heavens, and the other elements of this world, about the motion and orbit of the stars and even their size and relative positions, about the predictable eclipses of the sun and moon, the cycles of the years and the seasons, about the kinds of animals, shrubs, stones, and so forth, and this knowledge he holds to as being certain from reason and experience. Now, it is a disgraceful and dangerous thing for an infidel to hear a Christian, presumably giving the meaning of Holy Scripture, talking nonsense on these topics; and we should take all means to prevent such an embarrassing situation, in which people show up vast ignorance in a Christian and laugh it to scorn. The shame is not so much that an ignorant individual is derided, but that people outside the household of faith think our sacred writers held such opinions, and, to the great loss of those for whose salvation we toil, the writers of our Scripture are criticized and rejected as unlearned men. If they find a Christian mistaken in a field which they themselves know well and hear him maintaining his foolish opinions about our books, how are they going to believe those books in matters concerning the resurrection of the dead, the hope of eternal life, and the kingdom of heaven, when they think their pages are full of falsehoods and on facts which they themselves have learnt from experience and the light of reason? Reckless and incompetent expounders of Holy Scripture bring untold trouble and sorrow on their wiser brethren when they are caught in one of their mischievous false opinions and are taken to task by those who are not bound by the authority of our sacred books. For then, to defend their utterly foolish and obviously untrue statements, they will try to call upon Holy Scripture for proof and even recite from memory many passages which they think support their

position, although they understand neither what they say nor the things about which they make assertion.[292]

Those are wise words, from one of the wisest men who ever lived. It is good to have an informed opinion, but we must remember that our knowledge and understanding is finite and limited. We could be wrong about some of our cherished ideas. I have always tried to keep this in mind when studying theology and creationism. Some people believe this type of thinking is nothing more than a cop-out, but as a theologian I recognize that only God has perfect understanding of everything. The best thing an apologist can do is to recognize that we do not have all the answers. Sometimes saying, "I do not know for sure" is the best thing we can do.

Sometimes considering the ideas of the other side is easy – especially when they lack explanatory power! God's existence is a case in point. I can listen to an atheist all day long, but he or she will never persuade me from my Christian worldview. The atheist worldview is bankrupt. But when it comes to animal death before Adam, or 6,000 years versus 8,000 years for the age of the earth, or predestination versus free will, or any number of theological issues, I really must stop, listen, and think. There are times when we really need to abide by Augustine's advice. Never be afraid to listen to, and consider, the view of others. Maybe convinced unbelievers know a thing or two about science and even theology. The best theologians listen to others, even if they hold to a different position on origins or any number of other things. You might lose friends, and you will almost certainly never be offered to join certain ministries, but you will be honest with yourself – and you might learn something along the way. You will also be more like the noble Bereans, who tested everything by Scripture (Acts 17:11).

"Hold on a moment," my fellow young earth companions will say, "the age of the earth is an authority issue. God's words are the authority over all matters, and the Lord plainly told us he created in six days." I get that. I really do. But I have good friends, who are also mentors, that hold to the old earth position – and I know no one who reveres God's Word more than they do. They are not 'compromisers,' a term which my recent creationist brethren throw out far too often. For me, right or wrong I see the battle for the age of the earth as being an intramural debate. When it comes to telling unbelievers about God and his creation, the fact that the universe had a beginning is of far greater importance to me. The cosmological argument has far greater apologetic power than the issue of the earth's age. We will all do well to focus on the beginning, as opposed to the 'when' of creation.

---

[292] Augustine, *The Literal Meaning of Genesis* (Book 1, Chapter 18). http://inters.org/augustine-interpretating-sacred-scripture

The Days of Creation: 24-Hours or Long Ages?

There is another issue that is inseparably tied to the age of the creation: Are the days of creation 24-hours in length, or eons of time? As we have seen, the debate over the age of the earth and universe is nothing new. Yet, like the Bereans who studied God's Word in detail (Acts 17:11), we must examine for ourselves the clear meaning of 'day' in Genesis 1.

Recent creationists are adamant that the days of creation are six literal, 24-hour days, while old earth creationists insist that the days of creation were millions of years long.[293] But what does Scripture tell us regarding the days of creation? A typical concordance will point out that the Hebrew word for 'day,' *yom*, can have several meanings: A period of daylight as contrasted to nighttime, a 24-hour period, time in general, a specific point of time, or even a period of one year. Everyone agrees that *yom* has several meanings. No debate there.

John Oakes is the most knowledgeable, most dynamic apologist I know. He may not have the name recognition of William Lane Craig or Ravi Zacharias, but he is the best apologist I have ever seen or heard. I had the honor of studying apologetics under Dr. Oakes, when completing the Certificate in Christian Apologetics program through the Apologetics Research Society. Even though Dr. Oakes is an old earth creationist, he comments on the scriptural reasonableness of recent creationism:

> In answer to this point [Genesis 1], let it be said that six twenty-four hour days of creation is certainly a reasonable interpretation of Genesis 1. In fact, if it were not for what we know from science, it would at least appear to be the most obvious interpretation. However, one should bear in mind that the Hebrew word used for day here is *yom*. In the Old Testament, this word is variously translated 'day,' 'time,' 'forever,' 'age,' 'continuously,' 'today,' 'life,' and 'perpetually,' depending on the context. In fact, long before the scientific revolution, many of both Christian and Jewish theologians took a nonliteral-day interpretation of Genesis chapter one. For example, one could mention the Jewish theologian Philo, as well as the early Christian authors Justin Martyr, Irenaeus, Hippolytus, Clement of Alexandria, Origen, Lactantius, Eusebius and Augustine. This incomplete list proves that the idea of taking the six 'days' of creation to be ages rather than literal twenty-four hour periods is not necessarily a child of science.[294]

I agree with Dr. Oakes: If you remove science from the equation, the most

---

[293] Or, at the very least, there were millions of years between the days of creation.

[294] John Oakes, *Reasons for Belief: A Handbook of Christian Evidence* (Spring, TX: Illumination Publishers International, 2005), 217.

obvious interpretation of the Genesis creation account is six literal, 24-hour days. My question has always been this: Just how perfectly reliable, and trustworthy, is the science behind dating the creation, anyway? Just by asking this question, many people will automatically assume that I am either anti-science or a wild-eyed fundamentalist – or probably both. I assure you that I am neither. I believe that science is an ally in the search for truth, although it can take us only so far, and I do not consider myself even close to being a fundamentalist. (Chapters seven and eight will put that idea to rest.) In the previous section of this chapter, I attempted to make the point that there are many unknowns in science that have the potential to wholly change the age debate. As I said before, the creation could be old. But, then again, it could be much younger than academia and the media claim it is. Let us keep an open mind in this study.

Take time to read about the history of the age estimate for the earth. You will find that the numbers are all over the place, in every historical period. As Dr. Oakes pointed out, even in the early Church there were some who held to the possibility of a creation date far beyond that rendered through the biblical genealogies. Jonathan Sarfati, a recent creationist, confirms this point as well. He divides the early theologian-apologists into literal, figurative, and unclear categories concerning their interpretation of Genesis 1. Of the 25 names listed, nine held to a literal interpretation, four held to a figurative interpretation, and twelve were unclear as based upon their writings.[295] The age debate is old, indeed.

Table 6.2:

Early Church Leaders & Biblical Chronology

Literal Interpretation Days of Creation (24-Hour Days)

THEOPHILUS OF ANTIOCH	D. AD 185
METHODIUS	D. AD 311
LACTANTIUS	AD 240-320
VICTORINUS OF PETTAU	D. AD 304
EPHREM THE SYRIAN	AD 306-373
EPIPHANIUS OF SALAMIS	AD 315-403
BASIL OF CAESAREA	AD 329-379
CYRIL OF JERUSALEM	D. AD 387
AMBROSE OF MILAN	AD 339-397

---

[295] Jonathan Sarfati, *Refuting Compromise* (Green Forest, AR: Master Books, 2004), 121.

Figurative Interpretation Days of Creation
(Brief Moment or Long Periods)

PHILO	20 BC–AD 50
CLEMENT OF ALEXANDRIA	AD 150-215
ORIGEN	AD 185-253
AUGUSTINE	AD 354-430

Unclear Interpretation Days of Creation

JOSEPHUS	AD 37-100
JUSTIN MARTYR	AD 100-165
TATIAN THE SYRIAN	AD 110-180
IRENAEUS OF LYONS	AD 115-202
TERTULLIAN	AD 160-225
JULIUS AFRICANUS	AD 160-240
HIPPOLYTUS OF ROME	AD 170-236
EUSEBIUS OF CAESAREA	AD 263-339
GREGORY OF NYSSA	AD 330-394
GREGORY OF NAZIANZUS	AD 330-390
JOHN CHRYSOSTOM	AD 374-407
JEROME	AD 347-419

I highly recommend reading Richard Harter's "Changing Views of the History of the Earth," which may be found at Talk Origins Archive (talkorigins.org). This is not a website that makes recent creationists feel especially loved, but the article does reveal how the earth's estimated age has been debated throughout the centuries – and how it has increased steadily and tremendously over time. Although the author makes the point that the earth's age has increased due to scientific breakthroughs, I constantly asked myself as I read the article if the science is really that foolproof. What if there are major factors that we know nothing about? What if there are major factors that we assume to be true, but they are not? This has happened before in the history of science. Could the Flood have skewed our age estimates in a way that changes everything?

Recent creationists make the point – and it is a valid point – that a number and the phrase 'evening and morning' are used with each of the six days of creation (Genesis 1:5, 8, 13, 19, 23, 31). Outside of Genesis 1 *yom* is used with a number over 300 times, and every time it seems to refer to a 24-hour day. Why should we believe that Genesis 1 would be the exception? Outside of Genesis 1 *yom* is used with the words 'evening' or 'morning' 23 times, and every time the text seems to refer to a 24-hour day. Once again, why believe that Genesis 1 would be the exception? In Genesis 1:5 *yom* occurs in context with the word 'night.' Outside of Genesis 1, night is used with *yom* 53 times, and each time it seems to refer to a 24-hour day. Once again, why should we believe that Genesis 1 would be the exception? There are words in biblical Hebrew, *olam* and *qedem*, that are appropriate for communicating long periods of time. However, neither of these words are used in Genesis 1. In Exodus 20:11, where a number is used with days, it seems to refer to six 24-hour days.

James Barr (1924-2006) was a liberal professor of Hebrew and Old Testament at Oxford University. He did not believe in a literal reading of Genesis, nonetheless he admitted that as far as the language used in Genesis 1 is concerned, there is no professor of Hebrew or Old Testament Studies at any reputable university or seminary who would deny that the days of creation mean anything other than six 24-hour days. That was a huge admission for a liberal scholar. Likewise, Marcus Dods was a nineteenth-century liberal professor at New College in Edinburgh, Scotland. He correctly noted that if *yom* in the first chapters of Genesis does not mean a 24-hour day, then the overall interpretation of Scripture is hopeless. Does appealing to the admissions of liberal professors from the past prove that the days of creation were 24-hours long? No, but it does tell us that the issue is not dead and buried, as some skeptics and old earth creationists would like us to believe.

2 Peter 3:8 seems to equate one day with a thousand years. Therefore, skeptics of a recent creation claim that the days of Genesis 1 could be long periods of time. However, this verse is saying that, from God's perspective, time is meaningless since God is not locked into time as we are. A day is like a thousand years to God, because God is outside of time and not bound by it in any way. God is not limited by time, or any other created thing for that matter. What may seem like a long time to us is nothing to God. Using this verse to claim that a day in the Bible means a thousand years might also mean that Jonah was in the belly of the great fish three thousand years, or that Christ has not yet risen from the dead after two thousand years in the grave. This verse is really concerned with the end times, however, and reminds us that we are to be patient and not put a timeline on the return of Christ. Still, there are some skeptics of 24-hour days who are quick to use this verse in the context of creation, for their intended purpose.

Skeptics of recent creationism claim that six 24-hour days for creation limits God, whereas allowing God billions of years does not limit him in any way. However, this is not about limiting God, but rather it is about trusting what God plainly tells us he did. (Besides, who could possibly limit God anyway?) This is a 'trust God's Word' issue. Did God need eons of time? Of course not. God could have created in any timeframe he so desired – billions of years or even nanoseconds or less – but he chose six 24-hour days for a reason. Namely, the seven-day creation week serves as the template for our week: Six days of work followed by a Sabbath or 'Holy Day of Rest.' When you get right down to it, insisting on billions of years for creation limits God to man's timeframe.

There are some other issues with long ages for the days of creation. If the plants made on Day 3 were separated by millions of years from the birds (Day 5), nectar bats (Day 5), and insects (Day 6) necessary for their pollination, these plants could not have survived. Symbiotic relationships are when one type of plant or creature is dependent upon another type, such as the yucca plant and its associated moth. They were clearly designed with symbiosis in mind, so the origin of these plants and animals could not be separated by thousands or millions of years – or even months or weeks, for that matter.

Skeptics of recent creationism claim that since there is no 'evening and morning' for the seventh day of the creation week (Genesis 2:2), we must still be in the seventh day. For many people this means that none of the creation days are ordinary 24-hour days. However, God stated that he 'rested' from his work of creation on Day 7, not that he is 'resting.' The fact that he rested from his work of creation does not prevent him from continuing to rest from this work. God's work now is different: It is a work of sustaining his creation, not creating anything new (Colossians 1:16-17).

Regarding the seventh day of creation, *yom* is qualified by a number (Genesis 2:2-3), so the context still determines that it is a 24-hour day. Also, God blessed this seventh day and made it holy. In Genesis 3:17-19 we read of the curse on the earth because of sin, and Paul refers to this in Romans 8:22 as well. God would not call the seventh day 'holy and blessed' if he cursed the ground on this day. If we live in a sin-cursed earth – and we do – we are not currently in the seventh 'holy and blessed' day.

Some have argued that Hebrews implies that the seventh day is continuing today: "Now we who have believed enter that rest, just as God has said, "So I declared on oath in my anger, 'They shall never enter my rest.'" And yet his works have been finished since the creation of the world. For somewhere he has spoken about the seventh day in these words: "On the seventh day God rested from all his works.""[296] However, this passage reiterates that God rested (past tense) on the seventh day. If someone says on Monday that he rested on Saturday and is still resting, this does not mean that Friday continued through to Monday. It simply means that the person has continued to rest, over the course of a few days. Hebrews is not saying that the seventh day of the creation week has continued into today, merely that the rest from creating that God instituted on that day continues.

Skeptics of recent creationism claim that Adam could not have named all the animals on Day 6, as one day is not enough time to accomplish that task. Keep in mind, however, that Adam did not have to name all the animals, but only those that God brought to him. For example, Adam was commanded to name 'every

---

[296] Hebrews 4:3-4, NIV.

beast of the field' (Genesis 2:20), not 'beast of the earth' (Genesis 1:25). The phrase 'beast of the field' is almost certainly a subset of 'beast of the earth.' Adam did not have to name 'everything that creeps upon the earth' (Genesis 1:25) or any of the sea creatures. Additionally, the number of biblical 'kinds' would be much less than the number of species according to today's scientific classification.

When skeptics say that Adam could not name the animals in one day, they are really saying they could not do it – therefore, Adam could not have done it, either. However, the human brain has suffered for thousands of years from the effects of the fall. There is every reason to believe that before the degradation caused by sin entering the world, Adam's brain functioned at a higher capacity than do our brains today. That is not to say that we today are total boneheads, but we have suffered from the effects of the fall for a long time. (This is the effect of genetic entropy upon the brain and associated intellect.) Ancient man was truly intelligent and much more capable than most secular scientists give him credit for. (We will examine this issue in more detail momentarily.)

One last point of contention between young earth and old earth believers involves the order of creative events. Skeptics of recent creationism claim that there could not be day and night for the first three days of the creation, as Genesis 1 reveals that the sun was not created until Day 4. Evolutionism maintains that the sun came before the plants, with the sun playing a major role in the development of life on the earth. Keep in mind that the first three days of creation are written the same way as the last three days. If we let Genesis 1 speak clearly to us, all six days were 24-hours long. The sun is not necessarily needed for day and night. What is needed is a light source and a rotating earth. On the first day of creation, God made light (Genesis 1:3), and the phrase 'evening and morning' implies a rotating earth. Therefore, if we have light from one direction combined with a rotating earth, there can be day and night. Genesis 1:3 indicates that this light was created to temporarily provide day and night until God made the sun on Day 4. Revelation 21:23 tells us that one day the sun will not be needed because the glory of God will light the heavenly city. Perhaps God's glory was the light source for the first three days of creation.

Maybe God created light in this way to drive home the point that the sun is a created body, and not a divine entity as many ancient civilizations believed. The sun did not help create the earth, or life on the earth, in any way. Instead, the sun is God's created tool to rule the day that he had made (Genesis 1:16). Throughout history people such as the Egyptians have worshiped the sun. God warned the Israelites not to worship the sun as the pagan cultures around them did (Deuteronomy 4:19). They were to worship the God who made the sun, not the sun that was made by God. Evolutionary theories have always taught that the sun came before the earth, and that the sun's energy is responsible for life on this planet. It is interesting to contrast the ideas of modern cosmology with the writings of Theophilus, who lived in the second century: "On the fourth day the luminaries came into existence. Since God has foreknowledge, he understood the nonsense of the foolish philosophers who were going to say that the things produced on the earth came from the stars, so that they might set God aside. In order therefore that the truth might be demonstrated, plants and seeds came into existence before stars. For what comes into existence later cannot cause what is prior to it."[297] It appears that even in the second century the skeptics of the Bible were using the argument that life on the earth came from the sun, yet Theophilus was ready for them. There truly is "nothing new under the sun."[298]

Mankind: A Recent Creation?

Was mankind created or evolved? Scripture makes it clear: "But from the beginning of creation, 'God made them male and female.'"[299] Evolutionism maintains that mankind is very ancient, having evolved into his present form over the course of millions of years, while both young earth and old earth creationists are convinced that mankind is a 'recent addition' to the earth. The biblical genealogies do not render a vast age for humanity, but rather only thousands of years. Isaac Newton, in the eighteenth century, calculated the age of the earth to be 5,626 years, yet just twenty years later Benoit de Maillet confidently maintained that the earth was two billion years old.[300] Why the great difference among two men who lived so close in time to each other? It had everything to do with their worldview, not science. Newton, although an unorthodox Christian in some ways, nonetheless held the Bible in high regard while Maillet sought to destroy the foundations of Christianity. Maillet was highly influenced by Hinduism, which had become quite fashionable among some of the scholarly elite in Enlightenment-era France. Hinduism, as noted in chapter one, teaches that the universe is eternal, but marked by cycles of expansion and contraction, and Maillet accepted that idea wholeheartedly. (He therefore concluded that the earth had formed two billion years ago, during the current cycle of expansion and contraction.) Just twenty years later, George Louis Buffon taught that the earth was 75,000 years old.[301] The evolutionists were 'all over the map' concerning the age of the earth, but one thing was for certain among all of them: The biblical chronology of 6,000 or so years was much too inadequate for evolutionism. They required an ancient earth, and that is what they were going to present to the world. Their dedication to evolutionism drove their age estimates for the earth.

Yet people today assume that the earth and universe are ancient, on the order of billions of years. The reason they believe this can be summed up in one word: Science. Science has supposedly proven beyond a shadow of a doubt that the

---

[297] Theophilus, *To Autolycus* (Book 2, Chapter 15).
[298] Ecclesiastes 1:9, NIV.
[299] Mark 10:6, ESV.
[300] Herbert, 2.
[301] Ibid.

earth and universe are extremely old. But has science conclusively done that? Rarely does the non-scientist hear about scientific indicators for a young earth, but they are there. For example, DNA extracted from bacteria are claimed to be millions of years old, but DNA could not last more than thousands of years at best. This find might not prove a recent creation, but it does lend itself to that view quite well. Along the same line, many bones dated by secular science as being millions of years old are hardly mineralized, if at all, and some even contain still-pliable vasculature and blood cells. This would indicate that the bones are not that old, despite the common evolutionary claim that they must be very ancient.

The decay in the human genome is consistent with an origin for humanity only thousands of years ago. Secular-minded geneticists once taught that all women on the earth today descended from a small population of women in the distant past. These scientists collectively referred to this group as 'Mitochondrial Eve,' as if she were only one person. Likewise, these same scientists further concluded that all men alive on the planet today descended from a small group of men from long ago, collectively dubbed 'Y-Chromosome Adam.' Over time, something interesting happened: 'Mitochondrial Eve' and 'Y-Chromosome Adam' became more biblical. Instead of a small population of females or males, both figures became singular – and more recent. 'Mitochondrial Eve' has become just 'Eve,' and 'Y-Chromosome Adam' has become just 'Adam,' and based upon molecular dating they lived not all that long ago. This, of course, is consistent with the Genesis record. There are three other lines of evidence from human history that need to be considered as well.

First, there are not enough Stone Age skeletons. Evolutionary anthropologists say that the Stone Age lasted for at least 100,000 years, during which time the world population of Neanderthal and Cro-Magnon groups was roughly constant, between one and ten million. All that time they were burying their dead. By this scenario, they would have buried at least a few billion bodies. If the evolutionary time scale is correct, buried bones should be able to last for much longer than 100,000 years, so many of these Stone Age skeletons should be found. Yet only a few thousand have been excavated. This could easily be interpreted to mean that the Stone Age was much shorter than evolutionists think, possibly only hundreds of years in some areas, and not all that long ago.[302]

Second, history is much too short for evolutionism to be plausible. According to evolutionists, Stone Age man existed for 100,000 years before beginning to make written records only about 4,000-5,000 years ago. 'Prehistoric man' built megalithic monuments such as Stonehenge and the Great Pyramid, made beautiful cave paintings, and kept records of lunar phases.[303] Why would people

---

[302] Some tribes or groups would have lived a nomadic, 'Stone Age' lifestyle alongside more advanced civilizations. This is a trend that we observe even today.

[303] I have been to the Great Pyramid of Giza, and I can attest to the fact that it is mind-blowing! I have also examined very ancient cave paintings on the island of Sicily, and I can attest to their

wait a thousand centuries before using their advanced intellect to record history in written form? The biblical timescale seems much more likely in this regard.

Third, agriculture is too recent. The usual evolutionary picture has men existing as hunters and gatherers for 100,000 years during the Stone Age before discovering agriculture less than 10,000 years ago. Yet the archaeological evidence shows that Stone Age people were as intelligent as we are today. It is very improbable that none of the few billion Stone Age people, over many centuries, should discover that plants grow from seeds until only recently. It is more likely that mankind was without agriculture less than a few hundred years after the Noahic Flood, if at all. Evolutionists from the time of the Babylonians to the present have insisted upon ancient origins for mankind, yet that is not what the evidence demonstrates.

Adam or Ape?

Man has always been man, being created in God's image (Genesis 1:27). There was never a time when human beings were something less than they are now: The myth of the ape-man remains a myth. The australopithecines have been shown to be an extinct variety of ape, whereas both Neandertal man and *Homo erectus* are within the range of modern human anatomical variation. All the other so-called 'human missing links' can be explained as being either extinct varieties of apes or humans with archaic features.[304] There is no reason to believe that any of these fossil remains were somewhere between apes and humans. Keep in mind that of all the varieties of plants and animals (species and subspecies) that have ever existed on the earth, a high percentage are now extinct. It seems not only likely but highly logical that some of these extinct animals were primates – similar, yet different from, those in existence today. When looking at the fossil record, we can see skulls that appear to be either ape-like or human-like, but they have no living forms identical to them today. Does that make them a transitional form, or merely an extinct variety of ape or a human with extinct or rarely seen features today? (Such as the heavy brow ridge and sloping forehead of the Neandertals.)

Helmut Welke has explored the issue of human origins in depth. He comments about the current state of confusion in paleoanthropology:

> The journals of paleoanthropology now talk about a confusing
> 'tangled bush' of human ancestry, not a nice, clean diagram that
> is undisputed. The latest stylized drawings of this bush do not

---

magnificence as well. Ancient man was truly capable – scientifically, technologically, and artistically.

[304] An example of an archaic feature of man would be the skull shape of the Neandertals, who were human despite their minor anatomical differences.

even show a connection from Australopithecus afarensis (Lucy) to any of the above human fossils.[305]

Evolutionists should be questioning their favored theory, but instead many will be tempted to explain away the evidence against Darwinism. Fallen man will never consider creationism, unless first moved to do so by God (1 Corinthians 2:14-16).

Interestingly, the belief in ape-men existed well before Charles Darwin. The creationist Andrew Sibley notes that Lord Monboddo, a Scottish judge who was considered by some Scots to be the first modern evolutionist, held to the belief in ape-men:

> In the late eighteenth century Lord Monboddo was more determined to identify an evolutionary link between humans and apes. He practiced as a judge in Scotland, and was otherwise known as James Burnett (1714-1799). Monboddo was influenced by the Greek philosophers, and he argued that the Aristotelian great chain of being should extend from apes to man via a gradual process of evolution.[306]

Monboddo was clearly committed to proving the past existence of ape-men, just as Teilhard de Chardin was in the twentieth century and numerous evolutionists are today. The belief in ape-men even went back into antiquity: "Historical testimony suggests that popular mythology from antiquity, and its influence upon scientific discourse, had already primed European minds to accept the idea of evolution of man from apes."[307] Six centuries before Christ, Anaximander taught that human beings evolved from marine creatures, which had formed spontaneously from mud. Therefore, the idea that human beings descended from 'lower' creatures (even from mud) is nothing new. As Solomon told us three thousand years ago, "there is nothing new under the sun."[308] Yet here we are today, two and a half millennia after the time of Anaximander, and there is still no clear, undisputed evidence for ape-men. In fact, just the opposite is true: The case against ape-men is overwhelming.

But what about genetics? Is it possible that this branch of science could 'save the day' for evolutionists? Not even close. Helmut Welke explains how genetics has failed evolutionists:

> The fossils show no factual connection to the idea that humans evolved from an ape-like creature. Therefore, evolutionists hoped

---

[305] Email exchange between Helmut Welke and myself, November 25, 2020.
[306] Andrew Sibley, "Orang-Outang or *Homo Sylvestris*: Ape-Men Before Darwin." https://creation.com/apemen-before-darwin
[307] Ibid.
[308] Ecclesiastes 1:9, NIV.

that DNA would provide the needed evidence. The oft quoted 98.8% number for similarity between human and chimp DNA came from a review of an initial study that appeared in Nature magazine in 2005. But that was only the headline. When reading the article, and other related articles at the time, it becomes clear this was a preliminary number based on a limited study. (With some evolutionary assumptions added for good measure.) But the media and secular scientists took it as absolute truth. The study was based on less than 100 genes, that were already known to be similar. In other words, these evolutionary-minded scientists 'cherry-picked' their study. There are over 20,000 coding genes in our genome, and at least another 20,000 genes that are active controlling genes. These studies also ignored 'indels.' These are inserted or deleted nucleotides that are adjacent to similar runs of nucleotides.

Scientists should have known that indels would be an issue, because in a paper that was part of the Proceedings of the National Academy of Science the authors admitted that this 98.8% sequence identity drops to only 86.7% when taking into account the multiple insertions and deletions (indels). Note that while some insertions and deletions are small, some are up to 300 base pairs or more. This is quite significant, to say the least.[309]

Welke also noted that several evolutionists are now admitting that the 98.8% number is inaccurate, being closer to the low eighties. That may still seem like a high number to some people, but in terms of genetic variation that is a world of difference. He also reminded me about the work of Jeffrey Tomkins, who taught in the Department of Genetics and Biochemistry at Clemson University for a decade. Dr. Tomkins has published over fifty secular research papers in peer-reviewed scientific journals, and he has contributed seven chapters in scientific books in the areas of genetics, genomics, and proteomics. He was also the director for the Clemson Environmental Genomics Laboratory for three years. In other words, he knows his stuff! Now a staff scientist with the Institute for Creation Research, Dr. Tomkins confirms that we are nowhere near to being as closely related to the apes as secular scientists have long claimed.

There are a lot of problems with the idea of humans having evolved from an ape-like creature. A few of the key issues are enough to bury the idea once and for all. Take, for example, the difference in chromosome numbers between apes and humans:

Also, we know that humans have 23 pairs of chromosomes while all apes, chimps, and gorillas have 24 pairs. This is

---

[309] Email exchange between Helmut Welke and myself, November 25, 2020.

significant. Evolutionists speculated that two chimp chromosomes must have fused together into one human chromosome at some point in our evolutionary history. This was speculative, but they did identify two shorter chimp chromosomes and called them 2A and 2B. It was said that they fused together to form human chromosome number two.

But this idea has been studied further in the last few years and been shown to be a non-starter. The supposed fusion site has been shown to be a functioning gene, making it impossible to be a fusion site between telomeres (ends of chromosomes). There is also no sign of a non-functioning centromere on human chromosome number two. Every pair of chromosomes have a centromere site that helps align the pairs when they split for reproduction and then come back together. Human chromosome number two has only one functioning centromere and no sign of a second. The two chimp chromosome pairs (2A and 2B) would each have had a functioning centromere. Today the 'fusion story' has been totally debunked.

Humans and chimps are far from 98% genetically similar. The most recent genomic analyses show less than 85% genetic similarity, which is far from the alleged 98.8% that has been widely publicized for years. Additionally, there are still parts of the genome that cannot be aligned for comparison. Besides trying to compare 23 pairs (human) of chromosomes with 24 pairs (chimp), there are genes and sections of DNA that are significantly different in structure and are ignored when the similarity numbers are calculated. Clearly, humans did not evolve from an ape-like ancestor.[310]

Even if evolutionists could answer the problem of differing chromosome numbers – which is a monumental challenge – could there even be enough time for mutations and natural selection to change an ape-like creature into a human being? Evolutionists have long claimed that it took 3-3.5 million years for the australopithecines to change into a modern human being. Could genetics demonstrate that the change from a primitive ape to a modern human could be accomplished in that time frame? (Keep in mind that this is just the change from a primitive form of primate to a modern man. The evolution from a fish to a human – which was taught by Anaximander centuries before Christ, as well as by modern evolutionists who insist that the extinct fish Tiktaalik is a predecessor of amphibians, reptiles, mammals, and even human beings – would take considerably longer.) Could evolution 'work its magic' in three million years, bringing forth a man from an ape? Welke discusses the 'wait time' problem:

---

[310] Ibid.

Another issue is the 'wait time' problem. The question, "How long would it take ('wait time') for mutations and natural selection to evolve chimp DNA to human DNA?" is important. Modern genetic computer simulations have been done to answer this question. The results of a very sophisticated study appeared in the secular science journal Theoretical Biology and Medical Modeling in late 2015. To keep it simple, they simulated how long would it take for favorable mutations to become established or fixed in a population of hominids, of only two or five nucleotides in length. Remember, it is not just the case of a favorable mutation suddenly popping up. It must have some function and become established in the population, and not just die out again until it pops up again by chance.

The results were enlightening. (For creationists, that is!) To establish a string of two nucleotides required on average 84 million years. To establish a string of five nucleotides required on average two billion years. Using the most generous feasible parameter settings, the waiting time was consistently prohibitive. Even though evolutionists are adamant in their belief, they have no evidence for ape-to-human evolution. The wait time issue to establish favorable mutations in significant numbers is now shown to be scientifically impossible. Previously, evolutionists guessed 3-3.5 million years to go from Lucy to a modern human. This is nowhere close to the 84 million years demonstrated through the computer simulations described above. It is clear: We did not evolve from apes. Modern genetics demonstrates the impossibility of evolutionary thinking. That is good news for Bible-believing Christians, who always knew that God created mankind in his image (Genesis 1:27). You can trust your own direct reading of the Bible, without needing pagan beliefs – either ancient or modern – to interpret it.[311]

A quick internet search concerning the question, "how long has human evolution been going on," revealed that the estimate could be as high as eight million years for a primitive ape to evolve into a modern man. That is still a long way from 84 million years. According to evolutionists, the first primitive primates evolved 55 million years ago. However, that was long before the common ancestor of apes and man supposedly lived, and even that number is still a long way from 84 million years. (Plus, keep in mind this is based on the low number. A string of five nucleotides, rendering a two billion year wait time, is a far more likely scenario.) The fossil record and genetics fails to support human evolution. Although skeptics have long maintained that 'God was made

---

[311] Ibid.

in the mind of man,' the truth is instead biblical: God created man in his image (Genesis 1:27), and not all that long ago.

For further study on this topic, the DVD *Dismantled*, which can be purchased from either Answers in Genesis (answersingenesis.org) or Creation Ministries International (creation.com), is the best video available today. I highly recommend this resource if you want to learn more about human origins. *Contested Bones*, by John Sanford and Christopher Rupe, is also a must-have for further learning in this area.

The Advanced Capability of Ancient Man

If naturalistic evolutionism were true, ancient man was closer to a common ancestor with the apes than we are today, which would have made ancient man less intelligent than us so-called 'moderns.' If that were truly the case, we should expect to find that the further back we go in the archaeological record, the less advanced people seemed to be insofar as technological accomplishments. But is that what we find? Not at all. Instead, we find that the further back we go in the archaeological record, the more amazing are the accomplishments of ancient man. The advanced technology of ancient man points to an instantaneous creation of humanity (Genesis 1:26-27), in which the original people were created incredibly intelligent (Genesis 4:17b, 20-22) and, through the process of devolution as brought on by the fall of man (Genesis 3:17-19), became less technologically capable over time. Scientists today are still unsure how ancient man was able to construct the Great Pyramid at Giza, Egypt, as well as numerous other megalithic sites throughout the world.

Let us look at some of the amazing engineering accomplishments from antiquity. The Great Pyramid, on the Giza Plateau in Egypt, is the only one of the Seven Wonders of the Ancient World that still exists. Its' base spans thirteen acres, which is equivalent to five city blocks. It is 455' tall, making it the tallest building in the world before the Eiffel Tower in Paris was constructed. It consists of 2.5 million stones, each one weighing anywhere from 3-30 tons. Yet it is not just a lot of stones stacked upon each other: There are intricately designed passageways and chambers within its core. I had the great fortune in 2006 to visit the site, where I was not only awed by its' immense size – it truly is 'mind-blowing' – but I also had the chance to climb up the ascending passageway to the King's Chamber. It is incredibly designed, both inside and out.

Interestingly, the Great Pyramid may be described in Scripture: "In that day shall there be an altar to the Lord in the midst of the land of Egypt, and a pillar at the border thereof to the Lord, and it shall be for a sign and for a witness unto the Lord of Hosts."[312] The Great Pyramid stands at the ancient border that divided Lower Egypt from Upper Egypt, yet it is 'in the midst of' the land as a whole. It is simultaneously 'in the land' and 'at the border of the land.'

---

[312] Isaiah 19:19-20, NIV.

According to Josephus, it was built by Adam and Seth before the Flood. Recent creationists are convinced there is no reason to believe that it is anything but early post-Flood in origin, however, as the Noahic Flood was too destructive to allow any man-made structure to survive. Additionally, it is built upon sedimentary layers which are fossil-rich, so it gives every indication of being post-Flood in origin. On the other hand, many old earth creationists are open to the idea of pre-Flood origins for this structure, as they typically view the Flood as being both regional (non-global) in scope, and not necessarily as destructive as recent creationists maintain. (This is known as the 'Tranquil Flood' theory.) Either way, ancient man was supposed to be brutish and intellectually incapable at this earlier time in history, which does not square with what we see from observing this amazing structure.

No one knows who constructed the Trilithon Stones at Baalbek, Lebanon. The 'Stone of the Pregnant Woman' is the largest hewn stone in the world, weighing 1,200 tons. The other three stones at the site, which were laid out as a base or foundation for an unknown temple or building of some type, was well-utilized much later in history when the Romans built the Jupiter Baal Temple on top of them. Each of those smaller stones weigh approximately 750 tons – which is not small by any means! Yet ancient man was supposed to be a brute, only one-step away from a monkey in intellectual capacity. I think not!

Gobekli Tepe, in the Anatolia region of modern-day Turkey, contains the oldest known temple complex in the world, predating anything built by the Egyptians or even the Sumerians. There are twenty individual temples that have been discovered thus far, and the stones that comprise these temples are 40-60 tons each. The stones demonstrate intricately carved pictographs, mostly of animals. This site reveals an intricate level of design. Is Gobekli Tepe the work of a primitive hunting-gathering society? Not even close!

In addition to the sites already mentioned, there is Stonehenge on the Salisbury Plain in southern England, the incredible megalithic temples on the island of Malta, Puma Punku in Bolivia, Sacsayhuaman in Peru, the 44' long, 570-ton stone near the base of the Western Wall in Jerusalem, and the Carnac Stones in France. These are all sites that feature incredibly huge, yet intricately carved, stone structures. Mohenjo-Daro, at the border of India and Pakistan, is an ancient city which reveals the ingenuity of early man. It shows evidence of complex social engineering skills, including hot and cold running water. Finally, the Antikythera Mechanism, found off the coast of Greece in 1901, reveals a complex mechanical-analog 'computer' that was used to track the cycles of the solar system. It is believed to date to the second century BC, and it is likely of Greek or even Babylonian origin.

If molecules-to-man evolutionism were true, then how did ancient man – who, according to secular anthropologists, was brutish and dim-witted – design and build these sites or invent these complex 'computers'? Instead, the Genesis account of human origins – that is, man was created with an incredible degree of intelligence, which was progressively lost over time because of the effects of the fall – seems much more likely. In fact, it is the only scenario that makes sense of the archaeological data.

For further study on the topic of ancient man's abilities, I highly recommend Donald Chittick's *The Puzzle of Ancient Man: Evidence for Advanced Technology in Past Civilizations*, Bruce Malone's *Brilliant: Made in the Image of God*, and Don Landis' two-volume set *The Genius of Ancient Man: Evolution's Nightmare* and *The Secrets of Ancient Man: Revelations from the Ruins*. All four of these fascinating books detail the advanced science and technology that ancient man possessed, based upon archaeological findings.

Concluding Thoughts

We examined several points of contention between evolutionists and creationists in this chapter, mostly having to do with the 'dating game.' We saw that the creationist worldview holds up very well with the evidence from science, history, and logic. This is truly a great time to be a creationist!

We will now examine what I believe to be the greatest objection to the Christian faith today: The problem of suffering and evil. If you are a skeptic and this objection to the Christian faith is not number one on your list, I suspect it is not far behind. All people – believers and unbelievers alike – wrestle with this problem. It is truly the most formidable of skeptical objections for many people today, and it has been for a long time. As Christian believers we need to seriously wrestle with this issue – if even only for ourselves.

CHAPTER SEVEN

The Real Objection: Suffering & Evil

Some skeptics say that evolution is their primary reason for rejecting the God of the Bible, and I am sure that is true for some people. Other skeptics might say that the hypocrisy they have witnessed in the Church is enough to drive them away, or to keep them away if they were never part of the Church to begin with. The issue of hypocrisy is a real problem for skeptics, but it never bothered me as much as other issues. Even as a skeptic I could see that we are all hypocrites in some way, shape, or form – be it in religion or in any other way. Still other skeptics say their main reason for rejecting Christianity is that believers have an overly narrow view regarding other religions. During my unbelieving days I, too, felt that way. But I did not struggle with that objection as much as others. As a skeptic, all these issues and more got my attention. However, I was able to find satisfying answers to these issues. None of these skeptical claims were enough to keep me from investigating the faith with an open mind. But there was one issue that did rattle my cage: The problem of suffering and evil. It is, in my opinion, the one reason that skeptics give for rejecting the Christian faith that makes me say, "I get it." It is, in my opinion, the most formidable of all the objections to the Gospel message. All Christians need to wrestle with this problem, not only for themselves but for the benefit of those who they interact with.

Tied to the problem of suffering and evil is the belief that the God of the Bible is harsh and uncaring. Many skeptics believe that the God of the Bible is violent and unworthy of worship – if he even exists, that is. Charles Templeton, the former evangelist-turned-atheist, has said of the Old Testament God, "His justice is, by modern standards, outrageous...He is biased, querulous, vindictive, and jealous of his prerogatives."[313] What are Christians to make of this claim?

Thomas Jefferson, who was a deist and a Bible skeptic, described the God of the Bible as "cruel, vindictive, capricious and unjust."[314] Compared to the sentiments of his skeptical friend Thomas Paine, the author of the anti-Christian treatise *The Age of Reason*, Jefferson's assessment of God was tame. The situation has not improved since their time, so Christians must 'always be ready' (1 Peter 3:15) to address the issue of suffering and evil with the skeptics of the world.

Can believers offer valid explanations for the problem of suffering and evil, Canaanite genocide, and other supposed atrocities in the Bible? Can the explanations be both logical and scripturally sound? Believers have their work cut out for them, but there are answers. That is the beauty of apologetics.

---

[313] Lee Strobel, *The Case for Faith* (Grand Rapids, MI: Zondervan, 2000), 115.

[314] Ron Rhodes, *Answering the Objections of Atheists, Agnostics, & Skeptics* (Eugene, OR: Harvest House Publishers, 2006), 248.

In the next chapter we will examine the eternal fate of the unevangelized. This is an issue that is inseparably tied to the problem of suffering and evil, but it is so important to both believers and skeptics alike that it deserves a chapter of its own. If the only way to enter the presence of God when this life is over is by professing Christ as Lord and Savior while still alive on the earth, where does that leave those who never heard about Christ on this side of eternity? Are we to believe that these people will not be in Heaven because they did not profess Christ as their Savior, through no fault of their own? That certainly does not seem fair! We will also look at the reality and nature of Hell, an issue which sparks deep emotions in all people.

The skeptical concerns addressed in this chapter, and the next, are not exhaustive by any means. It is imperative that believers be ready to address all these issues and more, as skeptics have developed these objections into a comprehensive 'Case Against God.' This chapter, and the next, is for the purpose of demonstrating that the Christian believer has sound answers to these contentious difficulties of the faith.

Answering the Problem of Suffering & Evil

The problem of suffering and evil is a serious objection that cannot be skirted. Religious doubters are quick to say, "If God exists' he would have done something about the problem of suffering and evil. Since nothing has been done about it, God therefore does not exist." Other skeptics have decided that God does exist, but does not care about people, while other skeptics maintain that God is powerless to do anything about suffering and evil. Therefore, the skeptic believes that God either does not exist, does not care about us, or is powerless to resolve the problem. None of these options are good. Fortunately, there are answers to the problem of suffering and evil that are both logical and scripturally sound.

Christians do not deny that there is much suffering and evil in the world. Christian apologists commonly offer this approach to the problem of evil: If there is evil in the world, there must also be goodness because only through contrast can we recognize both good and evil. If there is both good and evil, there must be a moral law which distinguishes between the two. If there is moral law, there must be a moral Lawgiver. Only God could be great enough to ascribe moral law. Therefore, the moral law that distinguishes between good and evil is powerful evidence that God exists. It is not that God is imaginary, indifferent, or powerless, rather it is a lack of theological understanding that is sometimes the problem.

The word 'theodicy' comes up often when examining the problem of suffering and evil. Theodicy is derived from two Greek words, *theos* (God) and *dikei* (justice). Therefore, "a theodicy is a justification of the ways of God given all the evil and suffering that exists in the world."[315] Or, to say it another way, a

---

[315] Chad Meister, *Building Belief* (Grand Rapids, MI: Baker Books, 2006), 122.

theodicy is an explanation for the co-existence of God and evil. The term 'theodicy' is attributed to the philosopher Gottfried von Leibniz (1646-1716), in his book *The Theodicy*. But well before Leibniz, all people throughout history have wrestled with the problem of suffering and evil, attempting to either explain it away or to come to terms with it in some way.

The atheist attempts to explain the problem of suffering and evil by insisting that there is no God. Existentialism, at least in its atheistic form, is marked by the idea that 'life is absurd (or meaningless), so you might as well be happy and enjoy the ride for as long as it lasts.' As expected, this notion can lead to exaggerated forms of hedonism as a way of coping with the problem of suffering and evil, which is certainly not healthy.

Paul's philosophical audience in Athens (Acts 17:16-34) had their own ways of dealing with the problem of suffering and evil. The Epicureans sought to establish balance between pleasure and pain. If one is experiencing too much pain in life, the best thing that he or she can do is increase their amount of pleasure. However, this approach is nothing more than escapism, the same approach to pain and suffering that alcoholics and others with various addiction tendencies employ. Once again, not a healthy approach.

The Stoics, on the other hand, maintained that everything which takes place in the universe is the result of impersonal forces of nature over which we have no control. Therefore, the Stoic strove to control the only thing that we can control: Our response to the situation at hand. Through controlling the emotions to the point that nothing can disturb oneself, the Stoic sought to deal with the problem of evil through 'mind over matter.' In other words, if one does not mind, then nothing really matters. But does this idea really work? When life gets tough – and it will at some point – this approach of 'emotional super-control' will fail at some point. That is a given.

The deist accounts for the problem of suffering and evil by arguing that God is simply not concerned with the affairs of mankind. We are hurtling through space at thousands of miles per hour, and no one is in control. That thought is a little scary. The deist god is 'out there somewhere,' uninterested in our lives and our problems. Therefore, bad things just happen.

Finally, the Eastern religions teach that suffering and evil is an illusion. When one is doing fine, it may be possible to maintain this belief. But when pain and suffering enters' our lives – and it will – this idea of 'pain is just an illusion' will fall apart.

Table 7.1

Views on Suffering & Evil

VIEW #1:

ATHEISM – NO GOD, BAD THINGS 'JUST HAPPEN.'

VIEW #2:

EPICUREANISM & STOICISM – LIKE ATHEISM, BAD THINGS 'JUST HAPPEN.'

VIEW #3:

DEISM – BAD THINGS HAPPEN BECAUSE GOD IS NOT CONCERNED ABOUT US.

VIEW #4:

EASTERN METAPHYSICS – SUFFERING AND EVIL IS AN ILLUSION.

It is the Christian, however, who seems expected to shoulder the burden of providing a reasonable theodicy. The Christian knows that God exists (Romans 1:20), that life is not absurd (Ephesians 2:10), that there will be pain in this life (2 Corinthians 4:17), that our response to suffering and evil does matter (Ephesians 4:31-32), that God is extremely interested in the affairs of mankind (2 Timothy 3:16-17), that God cares for us deeply (John 3:16), and evil is not an illusion (John 20:27). Yet, accounting for these truths, the Christian must explain how a good and loving God can be reconciled with a world filled with suffering and evil. Fortunately, through knowledge of the doctrine of sin, we can do just that.[316] It may not be easy, but there are answers to this most formidable of all skeptical challenges.

The Free Will Theodicy

God created thinking beings with the ability to make choices for both good and evil (Genesis 2:15-17). Although it is preferable to make choices which promote goodness, as a moral world is more beneficial for everyone than an immoral world, thinking beings are free only if they are capable of decision-making without limitations. If God created beings without the ability to make choices, these beings would not truly be free – they would be nothing more than flesh-and-blood robots. Love is possible only when choice is possible: One being can love another being only when he or she chooses to love, as love cannot be forced. God could create beings capable only of making good choices which please him, but these beings would not be capable of truly loving God because

---

[316] The branch of theology that concerns itself with sin is known as 'hamartiology.' This term comes from the Greek words' *hamartia*, meaning 'missing the mark' or 'error,' and *logia*, meaning 'study.' Therefore, hamartiology is literally the study of 'missing the mark,' which is what sin causes us to do. Sin is an offense against God, through failing to keep God's holy standards.

they lack the ability to freely choose love. Without choice, there is no such thing as free will.

The free will theodicy, also known as the free will defense, rests on the foundation that the decisions which beings sometimes make are immoral and evil in nature, as they go against the goodness of God. As a result of these decisions evil is perpetuated in the world. Evil originated with the angels (Isaiah 14:12-14; Ezekiel 28:13-17; 2 Corinthians 11:14-15) then spread to mankind in part due to the deceptive nature of Satan upon the first humans (Genesis 3:1-7). History is a lesson in the fallen nature of mankind: It is a seemingly continuous story of pain and suffering, filled with immoral behavior directed against other people, animals, the environment, and even God himself.

Free will is an excellent explanation for the problem of moral evil in the world, but there is another type of evil that needs to be addressed as well. Much suffering is attributed to natural causes such as tornadoes, hurricanes, earthquakes, diseases, etcetera. These evils are not caused by the immoral actions of people, rather they are classified as 'natural evil.' At the time of the fall of mankind, when God cursed the earth (Genesis 3:17), the entire creation underwent a radical change for the worse (Romans 8:22). Since God pronounced his completed creation as 'very good' (Genesis 1:31), the creation obviously suffered from the rebellion of mankind. So, despite the claim of some theologians even natural evil is ultimately attributed to the fall of mankind. The nature of man is fallen, and the creation is cursed because of it. No one ever said life would be easy (Genesis 3:13-19).

The Soul-Making Theodicy

The soul-making theodicy was championed in modern times by the philosopher-theologian John Hick (1922-2012), although its roots may be traced all the way back to Irenaeus in the second century. This approach to theodicy maintains that evil exists for the purpose of exercising our ability to make proper moral choices and to teach us valuable lessons so as to 'build' our moral character and bring us more in line with how God intended us to be. Only by struggling against the evils of this world can we develop ourselves to be more like Christ. I like to think of the soul-making theodicy as the 'character development program' instituted by God.

The book of Job offers us an excellent example of the soul-making theodicy. Satan wished to make a point with God through Job, a man who was "blameless and upright; he feared God and shunned evil."[317] If Satan could crush Job, bringing him to his knees through pain and suffering, then Satan could prove to God that people love the Lord only as long as things are going well for them. God allowed Satan his 'test.' Yet, despite seemingly insurmountable odds, Job rises above his situation and shows God and Satan that man can love the Lord

---

[317] Job 1:1, NIV.

despite tragedy and extreme suffering.[318] In the end, Job's character is 'built up' beyond what it had previously been. He was a godly man before, but after his time of testing he became a superhero of the faith (Job 42:10-17).

Suffering and evil is a tough apologetic issue, and every believer must remember one thing of prime importance: When dialoguing with someone on this issue we cannot be 'just a theologian.' We must be both theologian and pastor. We need to be doctrinally sound, but even more important we must be compassionate. Let me give you an example of this point. In 2014 my dog Sadie died. Like many people, I view pets as being more than just pets – they are family. I was heartbroken, as Sadie was my 'four-legged daughter.' Shortly after her passing, an attendee in a class I was teaching tried to have a theological debate with me regarding animals in Heaven, and unlike me he was not open to that possibility. (A little sensitivity training may have been in order!) Although I typically love a healthy exchange of ideas, I was not ready for that dialogue on that day because I was still suffering. I can discuss the theological points of this topic now, years later, but not on that day. As much as I love a good theological discussion, I am always aware that there must be a proper time for it, and that time is not while someone is still suffering. So be considerate and compassionate of others. Making a theological point or winning a debate does not always matter. What does matter is showing Christ's love.

An Immoral God?
The Divine Command for Canaanite Destruction

When it comes to the problem of suffering and evil, skeptics are quick to point out the divine command to exterminate the Canaanites in the Old Testament. The books of Deuteronomy, Joshua, and 1 Samuel provide several passages which serve to stir up the emotional cry that the God of the Old Testament is genocidal. As an example, let us take a look at the book of Deuteronomy: "However, in the cities of the nations the LORD your God is giving you as an inheritance, do not leave alive anything that breathes. Completely destroy them – the Hittites, Amorites, Canaanites, Perizzites, Hivites and Jebusites – as the LORD your God has commanded you."[319] That should get the attention of anyone. As if that is not tough enough to read, there are many similar verses as well (Joshua 6:21; 8:24-26; 10:28; 10:40; 11:11; 11:14; 11:20-21; 1 Samuel 15:3).

These verses clearly demonstrate the command from God to kill all those who dwelled in the land previously promised to the Israelites, including women and children. In fact, the Canaanites' animals did not fare too well, either. For the believer this is often too difficult to think about, and for the skeptic these verses are a powerful line of evidence against the Bible – it is the problem of suffering

---

[318] Although this was not exactly a news flash for our omnipotent God!

[319] Deuteronomy 20:16-17, NIV. When I use the term 'Canaanite destruction' in this section, I am collectively referring to several tribes or people groups within the ancient land of Canaan, as outlined in these verses.

and evil in an early Old Testament setting. The believer must be prepared to address this specific issue, not only for the benefit of skeptics but, perhaps even more importantly, for the benefit of themselves.

It is necessary to begin with a brief examination of the Canaanite people. The Christian apologist Ron Rhodes states that, "God's command was issued not because God is cruel and vindictive, but because the Canaanites were so horrible, so evil, so oppressive, and so cancerous to society that – like a human cancer – the only option was complete removal."[320] Old Testament scholar Walter C. Kaiser, Jr. offers more details regarding the condition of the Canaanites: "When a people starts to burn their children in honor of their gods (Leviticus 18:21), practice sodomy, bestiality, and all sorts of loathsome vices (Leviticus 18:23, 24; 20:3), the land itself begins to 'vomit' them out as the body heaves under the load of internal poisons (Leviticus 18:25, 27-30)."[321] The Canaanites were a detestable people, burning their children as sacrifices to their false gods and committing grossly immoral sexual acts. Not a group of people that you would want to hang out with. (At least, I hope you would not want to hang out with them!)

The Christian apologist Clay Jones describes in more detail the practice of child sacrifice: "Molech was a Canaanite underworld deity represented as an upright, bull-headed idol with human body in whose belly a fire was stoked and in whose arms a child was placed that would be burnt to death."[322] Dr. Jones goes on to inform his reader of Plutarch's report from that time in history: "…the whole area before the statue was filled with a loud noise of flutes and drums so that the cries and wailing [from the sacrificed children] should not reach the ears of the people."[323] That is hard to even read, let alone think about.

History demonstrates time and again that one group of people can become an overwhelmingly bad influence upon another group of people. Could God allow the Israelites – his chosen people from which the Redeemer would eventually be brought forth – to be influenced by the Canaanites? If the Israelites had become like the Canaanites, would the promised Redeemer have come into the world? The world could in no way benefit from this depraved nation.

There is one major reason why God commanded the destruction of the Canaanite's but was not cruel or unjust for doing so: God's chosen people faced extinction at the hands of the Canaanites, and if the Israelites were wiped out the promise of the coming Redeemer would be wiped out as well. Morally, the Israelites could not integrate themselves into that depraved nation. If they became part of the Canaanite culture, adopting their immoral practices, they would have ceased to exist as a distinct nation. Extreme depravity will do that.

---

[320] Rhodes, 254.
[321] Ibid.
[322] Sean McDowell & Jonathan Morrow, *Is God Just a Human Invention?* (Grand Rapids, MI: Kregel Publications, 2010), 177.
[323] Ibid.

Yet if they refused to assimilate with the Canaanites, they would have been wiped out because of their 'high moral stance.' Immoral people never appreciate their moral neighbors. Either way, the Israelites would have gone the way of the dinosaur. Israel was fighting for her existence. (Unfortunately for the modern Israeli, the situation has not changed one iota.) Violence has been a sad fact of life in this fallen world, and the ancient Israelites were especially plagued by war and bloodshed.

God's plan for redeeming all humanity required the continuation of Israel, which would give rise to the promised Redeemer (Genesis 3:15). It was through Abraham that "all peoples on earth will be blessed,"[324] indicating that the Redeemer would come through the line of Abraham. God promised Abraham the land which later became inhabited by the Canaanites in the days of Moses and Joshua (Genesis 15:7). The land rightfully belonged to the people of Israel, and they were intent on reclaiming it. The coming Redeemer would arrive when the time was right (Galatians 4:4-5), and he would be born in the land promised by God. The Israelites had to reclaim the land of the Canaanites to fulfill messianic prophecy.

Unlike the Canaanites the Israelites did not fight wars for the sake of fighting. The Israelites fought only when necessary to preserve their existence. God made certain moral demands upon his chosen people, something that the other nations did not concern themselves with. Israel offered terms of peace before initiating war (Deuteronomy 20:10), and in most wars the Israelites were defending themselves and were not the aggressors (Exodus 17:8; Numbers 21:1, 21-32; 31:2-3, 16; Deuteronomy 3:1; Joshua 10:4). The enemies of Israel sought to eradicate them, yet the Israelites extended mercy to their enemies before each battle. This is a fact often missed by skeptics.

We must be aware that the command by God to destroy the Canaanites was unique and has not been repeated since. God is not in the business of ordering mass destruction. Also, keep in mind that Canaanite destruction had nothing to do with race or ethnicity. God ordered their destruction not because of their genome, but because of their extreme immorality. Therefore, despite the claim of skeptics this was not genocide, as it was not racially or ethnically based. Canaanite destruction was commanded by God for a specific purpose at that specific time in history.

One aspect of Canaanite destruction that is particularly troubling was the killing of children. Yet God demonstrated his compassion in that even though young Canaanite children would become casualties of war, it seems to many people that those under the 'age of accountability' would enter the presence of God after death. Isaiah 7:16 seems to hint at an age before a child is held morally accountable before God. If the Canaanite children of that generation grew into adulthood, they undoubtedly would follow in the depravity of their parents and become spiritually dead like those before them, entering eternity separated from

---

[324] Genesis 12:3, NIV.

God. Regardless of whether the 'age of accountability' is theologically true or not, we can say with confidence that God handles the death of young children with justice and mercy.

The issue of Canaanite destruction is difficult for skeptics and believers alike, but it may be effectively addressed through an examination of the doctrine of sin, knowledge of the Canaanite culture, and what God required of his chosen people in the Old Testament. At various times God demonstrates anger, justice, and even unimaginable mercy, grace, and love – as ultimately seen in the Savior himself, Jesus Christ. Unfortunately, Christ's coming into this world may have been thwarted in some way, shape, or form by the Canaanites.

A Plethora of Skeptical Objections

There are not just a few, but several objections to the faith that are posed by skeptics. The Old Testament has been a source of ammunition for many unbelievers, but as we will see Christians do have good answers to the challenges of the skeptical community.

How Could He? God's Command to Sacrifice Isaac

The account of God commanding Abraham to sacrifice his promised son Isaac is also a tough read. However, it is obvious from the context of Scripture that God never intended that this command be fulfilled, but rather it was a test of Abraham's faith: "Now I know that you fear God, because you have not withheld from me your son, your only son."[325] Some important points must be addressed concerning this account. God promised Abraham that, "I will make you into a great nation,"[326] and therefore Abraham could have confidence that Isaac, despite the impending sacrifice, would live on and have descendants. Otherwise, Abraham could not be the father of the great nation promised by God, since it was clear that Isaac was the child of promise (Genesis 17:21; 22:2).[327] Maybe Abraham suspected that God would not allow him to complete the sacrifice, or that God would raise Isaac from the dead. Either way, Abraham had confidence that the one true God would keep his promise and Isaac would live (Hebrews 11:17-19). Abraham said to his servants who had accompanied Isaac and himself to Moriah, "Stay here with the donkey while I and the boy go over there. We will worship and then we will come back to you."[328] That seems pretty confident to me.

In the ancient Near East, human sacrifice, including the sacrifice of children and infants, was commonly practiced. (As mentioned in the previous section, the Canaanites sacrificed an untold number of infants and small children to their

---

[325] Genesis 22:12, NIV.
[326] Genesis 12:3, NIV.
[327] Ishmael would also become a 'great nation' (Genesis 17:20), but he was not the child of promise.
[328] Genesis 22:5, NIV.

false god Molech.) The pagan cultures of the time sacrificed to their local false gods. The Old Testament contains several verses condemning this detestable practice (Leviticus 18:21; 20:2; Jeremiah 19:5; Ezekiel 20:30-31; 23:36-39). Even though Abraham lived before the time that these verses were written, it is clear that this command to avoid human sacrifice has always been in place (Hebrews 13:8). Human sacrifice to the false gods is not the same as a test of faith by the one true God – and Abraham knew the one true God, who would not command him to do anything for the purpose of evil.

Abraham's obedience to God is a foreshadowing of the sacrifice which God the Father would eventually endure to completion through his Son, the promised Redeemer. Abraham, through his willingness to sacrifice Isaac, was faithful to God. In turn, God's sacrifice of his Son demonstrates his faithfulness to mankind. That is the process of mutual love that believers should always keep before them.

Although I hope I am never tested by God in similar fashion, or as Job was tested (Job 1:8-12; 2:3-6), it is reassuring to know that God is in control of everything (Psalm 115:3; Proverbs 16:9; 19:21; Isaiah 45:6-7; Romans 8:28; Ephesians 1:11; 1 Timothy 6:15). But, despite this, it is ultimately reassuring to know that, "God so loved the world that he gave his one and only Son, that whoever believes in him shall not perish but have eternal life."[329] Because of God's great love for mankind, we can be confident that he will care for us in all ways, even if we do not fully understand why he asks certain things of us.

God Creates Disaster?

Isaiah 45:7 states, "I form the light and create darkness, I bring prosperity and create disaster; I, the LORD, do all these things."[330] Disaster does not equate to moral evil, however. Hebrew linguists state that the word disaster, as used in this verse, need not have any moral connotations. Scripture is clear that God is morally perfect (Deuteronomy 32:4; Matthew 5:48) and does not sin (Hebrews 6:18). In fact, God's absolute justice (Psalm 145:17; Revelation 15:3-4) demands that he punish sin. God may use disaster, such as the forces of nature or plagues, to bring about desired results. An example of this is the plagues inflicted on the Egyptians immediately prior to the Israelite exodus from Egypt (Exodus 7:14-11:10).

Regarding Isaiah 45:7, "The best answer might rest in the image of God as a loving parent, who offers care but insists on obedience."[331] Only as parents do we more fully comprehend the necessity of both loving and disciplining our children. It is not only possible to both love and discipline our children at the same time, but as parents we must do so.

---

[329] John 3:16, NIV.
[330] Isaiah 45:7, NIV.
[331] NIV Quest Study Bible, 1046.

Did God Sanction the Sacrifice of Jephthah's Daughter?

This skeptical objection is based upon Judges 11:30-40. This has been a difficult passage of Scripture for biblical scholars, and varying explanations have been proposed over time. At the time of the Judges, the people were in a period of history when everyone did what was right in his or her own eyes: "In those days Israel had no king; everyone did as he saw fit."[332] It may be that Jephthah literally sacrificed his daughter. He may have believed this to be the correct thing to do ('as he saw fit'), due to his sacred vow with God. However, God had commanded in the days of Moses that human sacrifice not be performed (Leviticus 18:21; 20:2-5; Deuteronomy 12:31; 18:10), so if Jephthah did sacrifice his daughter, he was wrong for having done so. Just because the Bible records an action does not mean that God condones that action. For example, God does not agree with the words or actions of Satan, but nonetheless the Bible does record the thoughts and deeds of the Devil.

Another explanation for this contentious passage is that Jephthah offered up his daughter as a 'living sacrifice,' consecrating her for service at the tabernacle for the rest of her life. If this were the case, then Jephthah's daughter would have been celibate the rest of her life, a serious matter in ancient Israel where women were encouraged to have children and continue their father's lineage. This may account for the fact that Jephthah's daughter went off into the hills for two months to weep with her friends, to mourn for not being able to marry and have children. In neither explanation did God command Jephthah to do anything to his daughter. Jephthah is the one who took matters into his own hands, whatever that may have been.

Did God Sanction Slavery?

The 'New Atheist' Sam Harris is very quick to use this common objection to Christianity: "Consult the Bible and you will discover that the creator of the universe clearly expects us to keep slaves."[333] This issue has been one of the most severe allegations that skeptics have utilized in their case against the character of God.

In Genesis 1:27 we learn that God created all people 'in his image,' and as a result God does not condone slavery, which is the act of one people exercising an unfair rule and authority over another people, usually to exact labor from them. Paul states, "From one man he made every nation of men, that they should inhabit the whole earth; and he determined the times set for them and the exact places where they should live."[334] The Creator has made humanity in his image, and all people literally share a common ancestor,[335] with no divine favoritism

---

[332] Judges 21:25, NIV.

[333] McDowell & Morrow, 148.

[334] Acts 17:26, NIV.

[335] Although some adhere to the belief that Adam and Eve are the ancestors of the Semitic people only, subscribing to a form of theistic evolutionism which makes a distinction between the people

being shown for one people group over everyone else. Even God's chosen people, Israel, were chosen not as God's favorite people, but rather as God's ambassador's to the world, to tell the world about the one true God (Genesis 12:1-3; Exodus 9:16; Psalm 67; 96; Acts 1:8; Acts 13:47). The Bible records the fact that slavery did occur, and through a surface reading it does seem that God is fine with the institution of slavery. But, as we have done throughout this book, we are going to be like the Bereans and go well beyond a mere surface reading (Acts 17:10-11).

Skeptics are correct to point out these verses, as they are potentially problematic for Christians – they must be acknowledged as being contentious. Slavery was a part of the Semitic cultures of the ancient Near East, and the Israelites were certainly participating in this practice – despite having been on the losing side of it at one time. However, upon examination of the relevant passages it is clear that the Israelites were commanded by God to treat servants and slaves fairly. It is clear from history that no other nation or group of people besides the Israelites adhered to a divine command to treat their slaves fairly (Exodus 1:6-14). The Christian ethicist Paul Copan notes that "slavery in biblical times was different from slavery in the old South,"[336] which is usually the example of slavery that comes to mind for most people today. During the first century, when Paul was writing his epistles, well over half of Rome's population were slaves in some way, shape, or form. Slaves had many rights during this time in the Roman Empire: The ability to start their own business, the possibility of earning enough money to purchase their freedom (manumission), and the right to own property (peculium). It must also be realized that many slaves held positions of power and authority. Therefore, slavery in ancient Rome was not nearly the same as slavery in the American South, or like the slavery the Hebrews endured at the hands of the Egyptians. (Slavery is never good, but there have been different levels of harshness throughout slavery's sordid history.) The most difficult verses concerning slavery are Leviticus 25:44-46, in which God states that the slaves of the Israelites must come from the other nations. Although God states that the Israelites must not rule over their own people ruthlessly, he does not say the same when discussing the slaves from other nations. However, in light of God's other commands to treat servants and slaves fairly, and in memory of their own time in bondage in Egypt, it may be reasonably assumed that the Israelites were expected to treat the slaves from the other nations fairly as well. Copan notes that slavery, "was mitigated, limited, and controlled in the law of Moses."[337]

Paul wrote, "There is neither Jew nor Greek, slave nor free, male nor female,

---

of the ancient Near East – specifically the Semites – and the rest of humanity. However, the best evidence from biology and anthropology supports a common ancestor for all human beings, as clearly described in Genesis.

[336] Paul Copan, *That's Just Your Interpretation* (Grand Rapids, MI: Baker Books, 2001), 172.
[337] Ibid.

for you are all one in Christ Jesus."[338] The New Testament reveals that we are all equal before God. Paul also urges, "Slaves, obey your earthly masters with respect and fear, and with sincerity of heart, just as you would obey Christ."[339] Paul is not condoning slavery, but rather noting that since slavery was a cold, hard fact of his day, slaves needed to obey their earthly masters just as they would obey Christ. It may be that the biblical writers did not condemn slavery simply because, "social reform was secondary to certain internal, attitudinal transformations [salvation]."[340] Perhaps in the end, all we can really say is that the biblical writers neither approved of or endorsed slavery, but realized that it was part and parcel of life in a fallen world so they attempted to regulate it as best they could.

The matter of slavery in the Bible is a difficult issue for the Christian believer, but it must be emphasized that God has created all people in his image (Genesis 1:27), we are meant to be our 'brother's keeper' (Genesis 4:9), we are equal descendants of our original parents, being 'of one blood' (Acts 17:26), and we are equal in Christ (Galatians 3:28). Additionally, history furnishes us with knowledge regarding the culture and practices of the ancient Near East, shedding further light on the matter at hand. God demanded that the Israelites – who, unfortunately, participated in slavery – treated their slaves fairly. No other nation had this divine command.

Did God Harden Pharoah's Heart?

In Exodus 4:21, we read that God hardened the heart of Pharoah. On the surface, this seems to be a horrible act by God: How could the Lord harden the heart of anyone, thereby denying that person the ability to make proper moral choices which could eventually serve God and his followers for good? However, the biblical record states that Pharoah hardened his own heart (Exodus 7:13-14, 22; 8:15, 19, 32; 9:7, 34-35; 13:15) before God hardened his (Exodus 4:21; 7:3; 9:12; 10:1, 20, 27; 11:10; 14:4, 8, 17). Sadly, we are just like Pharoah: In the end, people ultimately harden their own hearts. Paul, in Romans 1:24-28, describes a similar situation in which God gives people over to their own choices:

> Therefore God gave them over in the sinful desires of their hearts to sexual impurity for the degrading of their bodies with one another. They exchanged the truth about God for a lie, and worshiped and served created things rather than the Creator – who is forever praised. Amen. Because of this, God gave them over to shameful lusts. Even their women exchanged natural sexual relations for unnatural ones. In the same way the men also abandoned natural relations with women and were inflamed with

---

[338] Galatians 3:28, NIV.
[339] Ephesians 6:5, NIV.
[340] Copan, 175.

lust for one another. Men committed shameful acts with other men, and received in themselves the due penalty for their error. Furthermore, just as they did not think it worthwhile to retain the knowledge of God, so God gave them over to a depraved mind, so that they do what ought not to be done.[341]

The question which really must be asked is this: Is God hardening hearts randomly, or do people first make the choice to harden their own hearts and God merely helps them along after a certain amount of time, ultimately using their spiritual stubbornness to serve a greater purpose? This seems to be the case regarding Pharoah. We must learn to trust God and live according to his will.

Did God Cause Prophets to Lie?

This accusation comes from 2 Chronicles 18:20-21. However, it is clear from other verses in Scripture that God detests lying (Exodus 20:16; Psalm 59:12; Proverbs 12:22). In fact, God cannot lie (Numbers 23:19). So, why would God cause prophets to lie? Ron Rhodes reminds us that, "We must make a distinction, however, between what God causes and what He allows…God permitted the activity of a lying spirit, but He did not cause it. Therefore, God's character cannot be impugned."[342] Rhodes makes a valid point by distinguishing between allowance and causation. God grants us free will, and although God may not like the choices which we sometimes make, free will is nonetheless a gift from God which we exercise constantly. "In the final analysis, God did not deceive Ahab. He gave Ahab the choice between believing a lie and believing the truth. But Ahab already had his mind made up and chose what he wanted to believe."[343] Once again, we will do well to remember that the Bible is often descriptive, rather than prescriptive, concerning the events that are recorded in its pages.

Did God Create the Wicked So He Could Bring Disaster Upon Them?

Proverbs 16:4 seems to indicate that God is responsible for setting up some of us to fail, and fail horribly at that: "The LORD works out everything for his own ends – even the wicked for a day of disaster."[344] This verse does not mean that God created certain people wicked – that is to say, wicked beyond our inherently fallen nature – just so God could destroy them in some manner, or send them to Hell. That would make God evil, and it is clear from Scripture that God is anything but evil. God is holy, loving, and just. In fact, God is not willing that any should perish: "The Lord is not slow in keeping his promise, as some understand slowness. He is patient with you, not wanting anyone to perish, but everyone to come to repentance."[345] It is also clear that God wants all people to

---

[341] Romans 1:24-28, NIV.
[342] Rhodes, 253.
[343] NIV Quest Study Bible, 626.
[344] Proverbs 16:4, NIV.
[345] 2 Peter 3:9, NIV.

be saved: "This is good, and pleases God our Savior, who wants all men to be saved and to come to a knowledge of the truth."[346] John 3:16, perhaps the most well-known verse in all Scripture, makes it abundantly clear that God loves this world: "For God so loved the world that he gave his one and only Son, that whoever believes in him shall not perish but have eternal life."[347] This is clearly not the description of a God who is cruel and vindictive. God uses the free will actions of people to orchestrate his eternal plan, of which we know something about based upon Scripture, but certainly not all of his plans are known to us (Deuteronomy 29:29; Isaiah 55:8-9). The wicked have chosen to be wicked, and God uses them just as he uses everyone to play a part in his 'Grand Design' for human history.

Unanswered Prayer: A Serious Challenge to Faith

As a believer, one thing that challenges my faith tremendously is unanswered prayer. Or, more accurately, prayers that are not answered the way I hoped they would be. I am convinced that God hears our prayers, and I am convinced that God cares about us, but sometimes it is hard to understand why things happen the way they do.

If you read the dedication page to this book, you will see that my brother-in-law Frank died during the time I was writing this book. Over twenty-five years ago, Frank developed an aggressive cancer that almost took his life. Many people prayed for him during this time. I prayed earnestly, for months on end. It was touch-and-go at times, but Frank came through it and beat cancer. Our prayers had been answered. Frank was able to see his two boys grow up and attend college, establish themselves in their chosen careers, and he attended the wedding of his eldest son. In the years before his passing Frank got to experience one of the greatest joys that anyone will ever know: Grandchildren. His two grandsons were the highlight of his final years. Long ago we prayed for him to beat cancer, and he did. But there was a price to pay for those cancer treatments.

Between the chemotherapy and the radiation treatments, his heart barely functioned later in his life. His last few years were up-and-down, and he experienced one difficulty after another. He needed a heart transplant, but he was also aware that most people die before they receive a new heart. Unfortunately, this happened one day after his forty-third anniversary. This was certainly not what we had hoped, and prayed, for.

During Frank's final months, my mother experienced a cascade of medical problems that would have taken the life of most people her age. She was never considered strong enough to survive a bladder surgery that she needed years ago, but during this time she underwent an emergency cardiac ablation with a pacemaker placement, and barely two months later she underwent surgery for a broken hip that happened during an episode of pneumonia. We have no idea how

---

[346] 1 Timothy 2:4, NIV.
[347] John 3:16, NIV.

she survived any of this, and at times we did not think she would. Thank God she did.

During this difficult time, I underwent back-to-back cardiac ablation procedures for a heart arrhythmia that I developed in the previous year. Between the procedures, which were only two weeks apart, I could have died myself. (One night, while alone, I thought I was going to die. Yet I was able to call 9-1-1, unlock the front door, and sit on the couch while waiting for EMS to arrive, all while in ventricular tachycardia. This is a potentially deadly heart arrhythmia. It is amazing what we can do when we have to!)

By the way, did I mention that all of this happened in 2020? The year that stunk for everyone on the planet. Between the COVID-19 pandemic, racial tensions, a failed economy, a mind-numbing presidential race, and a host of other problems, I have no idea how anyone made it. (As I write this, it is only mid-October, so maybe I should not get too confident!) I, and the people in my family, prayed a lot this year. I suspect that even those of a secular disposition spoke to God a time or two.

Some prayers were answered this year, but not others. Frank's passing was by far the worst thing. Nothing else even came close, and that is saying a lot in this crazy year. Yet both my mother and I have so far done well, and at this point everyone in my family has remained COVID-free. I pray that continues.

Many people prayed for Frank to get a new heart and do well. But it did not work out that way. Prayer worked before, when Frank had cancer, but it did not work this time.[348] I am thankful, however, for the many years that Frank had between his cancer treatments and his passing. But I know that many of us are left wondering why God did not answer our prayers this time. Some might wonder if God even exists.

At this point in our lives, individually and as a family, we do not know how Frank's passing will impact us. We may only know that answer in the end, as we look in the rearview mirror of our lives. God exists, however. I know this from science and Scripture. God's Word reveals something of profound importance: "We know that for those who love God all things work together for good, for those who are called according to his purpose."[349] Believers often quote this verse during times of loss or difficulty, but I never do. I feel like I do not know enough about God's purposes for allowing certain things to happen. I also fail to see how any of this can be good. At least now. (When I am in eternity, I might look back and be able to make sense of it. But then again, I probably will not even bother to look back at that point.) I am confident, however, that near the

---

[348] Of course, skeptics who read this will quickly conclude that Frank's triumph over cancer years ago was merely serendipitous. In other words, God had nothing to do with it: The body, which is an amazing machine, simply healed itself. After all I have written in this book, however, I cannot come to that conclusion. If God created the universe and everything in it, including people – which he did – then it makes perfect sense that he can 'hear' (audibly or otherwise) our prayers, petitions, and requests.

[349] Romans 8:28, ESV.

end of our lives we will all reflect on our past and have a better idea why certain things happened the way they did. Some prayers are not answered the way we would like them to be, but that does not mean that God is non-existent, or uncaring, or powerless to help us. It simply means that he knows more about how it will all work out in the end then we do.

For now, we can say with certainty that in this life things are not always fair – at least as we believe we understand fairness. But then again, no one ever said that life would be fair. People have an inherent sense of justice: We believe we know what should, and should not, be allowed to happen. For example, I never smoked, drank little-to-no alcohol, exercised a lot, and led an active life. I knew my genetic history on both sides of my parentage was bad, in terms of cardiac disease, but that would probably affect me someday in the future, well past retirement. Then, on December 28, 2018 I suffered a heart attack. Technically, it was a ventricular tachycardia event, which is an especially serious rhythm disorder. It was a bad situation: I was shocked three times in about an eight-hour period, and statistically I was lucky to have made it. The problem is, I did not feel lucky! I was the guy that always worked out and took care of himself. But I got cocky: I thought I had more control over my life than I did. After this close brush with death, I read these words from James, the earthly brother of Jesus:

> Now listen, you who say, "Today or tomorrow we will go to this or that city, spend a year there, carry on business and make money." Why, you do not even know what will happen tomorrow. What is your life? You are a mist that appears for a little while and then vanishes. Instead, you ought to say, "If it is the Lord's will, we will live and do this or that."[350]

I thought I had it figured out, but I was wrong. God is in control, not me. Although I would have preferred another way to learn, God taught me a lot in the two years since my cardiac event. Life is a gift from God, and I have everything to be thankful for. I complain like most everyone does, but I have no reason to do so.

Frank passed way too soon, in my mind. My best friend passed at only 50 years of age, after a brutal fight with cancer. He left behind a young family. Was that fair? Not in my mind. Pastor Kirk Kendall was one of my mentors. He was a great man who taught me a lot about leadership. He did not have a chance to retire and enjoy fishing full-time, which he loved. Was that fair? Not in my mind. Jhan Moskowitz, one of the founding members of Jews for Jesus and a friend of mine, died after an accidental fall. He had a lot of ministry left in him. Was that fair? Not in my mind.

But guess what? My mind is limited. Big time limited! There is a lot I do not understand, and even in eternity there will likely still be things that I will not

---

[350] James 4:13-15, NIV.

understand. But God knows everything. He knows the beginning from the end, and he knows every detail of every life that has ever been lived. I – make that 'we' – need to learn to trust God, even when we do not understand things. But there will be times, especially during suffering, when it is hard to trust God.

Evolutionism: A Cause of Suffering & Evil

Does evolutionism help or hurt the world? I contend that the philosophy of evolutionism has been a cause of pain and suffering in this world. A philosophical or religious doctrine can be measured in terms of its worth by looking at its so-called 'fruits.' Just what are the fruits of evolutionism throughout history? For starters, racism, class suppression, and slavery. Human sacrifices, gladiatorial games, and other types of mass slaughter were common in evolution-influenced cultures. Abortion, infanticide, misogyny, and ultimately no concern for others was rampant. In short, the devaluing of all life. And why not? If life on the earth is seen as being nothing more than a cosmic accident, and people arose from lower life forms – and ultimately are nothing more than 'rearranged pond scum' – then what difference does it make in how people and animals are treated? Many people are convinced that, since there is no personal God to answer to when this life is over, we can get away with a multitude of things in this life. Life is harsh, so why stick around for long? Maybe we should just end our lives prematurely, and maybe killing someone else is doing them a favor. That is the logical conclusion of a godless, evolutionary worldview taken to its extreme.

Contrast that with biblical creationism. An all-powerful God designed this universe and everything in it with care and precision. Human beings were even created in God's image (Genesis 1:27), designed to know that they will live forever (Ecclesiastes 3:11) and equipped to inherently know right from wrong (Romans 2:14-15). People were intended by God to be good stewards of the creation (Genesis 1:28-30) and to be their 'brother's keeper' (Genesis 4:9). What are the fruits of that worldview? Not racism, not class suppression, not slavery, not public executions, not abortion, not infanticide, and not misogyny. In other words, the opposite of the fruits of evolutionism.

The Code of Hammurabi, dated to 1754 BC, summarized the harsh code of conduct for the ancient world: 'An eye for an eye, and a tooth for a tooth.' That is a reasonable philosophy for one who believes in an evolving world that is 'red in tooth and claw.' In the movie *Conan the Barbarian*, the main character Conan is asked the question, "What is best in life?" His answer was, not surprisingly, barbaric: "Crush your enemies, see them driven before you, and hear the lamentations [cries] of their women." That was the worldview philosophy of the ancient world. It was a tough place – definitely not for the faint of heart. The twenty-first century is not necessarily a great place to be in, either, but it beats life in the ancient world.

Eventually a different code of conduct entered the world, put forth by one who not only spoke truth, but was the embodiment of truth itself (John 14:6):

> You have heard that it was said, 'Eye for eye, and tooth for tooth.' But I tell you, do not resist an evil person. If anyone slaps you on the right cheek, turn to them the other cheek also. And if anyone wants to sue you and take your shirt, hand over your coat as well. If anyone forces you to go one mile, go with them two miles. Give to the one who asks you, and do not turn away from the one who wants to borrow from you. You have heard that it was said, 'Love your neighbor and hate your enemy.' But I tell you, love your enemies and pray for those who persecute you.[351]

This is a code of conduct based upon a creation worldview, where people are viewed as bearing God's image and are of immeasurable worth. This is a code of conduct based upon the God of love.

When it comes to contrasting the fruits of evolutionism with the fruits of creationism, Paul said it best:

> The acts of the flesh [fruits of evolutionism] are obvious: sexual immorality, impurity and debauchery; idolatry and witchcraft; hatred, discord, jealousy, fits of rage, selfish ambition, dissensions, factions and envy; drunkenness, orgies, and the like. I warn you, as I did before, that those who live like this will not inherit the kingdom of God. But the fruit of the Spirit [fruits of creationism] is love, joy, peace, forbearance, kindness, goodness, faithfulness, gentleness, and self-control. Against such things there is no law. Those who belong to Christ Jesus have crucified the flesh with its passions and desires. Since we live by the Spirit, let us keep in step with the Spirit. Let us not become conceited, provoking, and envying each other.[352]

Although Paul is discussing the fruits of the fallen nature versus the fruits of the regenerated man in the above verses, the broad concept is the same when applied to evolutionism and creationism. Has evolutionism benefitted mankind? You decide.

Concluding Thoughts

For me, the problem of suffering and evil is the most serious objection against the Christian faith today, which is why I avoid harshly criticizing atheists and skeptics who thoughtfully wrestle with this issue. I may not agree with their worldview, but I do understand why so many skeptics avoid the life of faith

---

[351] Matthew 5:38-44, NIV.
[352] Galatians 5:19-26, NIV.

because of so much suffering and evil in the world. That is how I felt about it when I was a skeptic many years ago, and I still recognize the problem of suffering and evil as being a serious challenge today. (For me personally, it has always been the greatest challenge to biblical faith.) However, when compared to the other worldviews, the biblical approach does offer the most satisfying answers to the co-existence of a good God and an evil world.

It is clear, after thoroughly studying both testaments, that the God of the Old Testament is also the God of the New Testament; there is, after all, only one God, and he never changes (Malachi 3:6). God judges in both the Old and the New Testaments, and he also demonstrates mercy in both testaments as well. God displayed mercy at the time of the fall of man, when, instead of destroying our original parents, he instead promised the future Redeemer (Genesis 3:15). This involved the ultimate act of self-sacrifice. Additionally, although God was once "grieved that he had made man on the earth,"[353] he nonetheless provided for the continuation of mankind through his servant Noah (Genesis 6:9-22), and God's provision of the covenants demonstrate love and grace (Genesis 12:1-3; 2 Samuel 7:8-16).[354]

Jesus, undoubtedly the greatest moral teacher in all of history, exemplified love. However, he also denounced those who conducted themselves poorly and mistreated others. His interaction among the money changers in the Temple was not the loving Jesus we often think of, but every time he acted harshly toward people it was because the situation required it.

The list of supposed atrocities that skeptics level against Christianity extends well beyond what is covered in this chapter. It is clear that no one can have all the answers to the skeptical claims which challenge the Christian faith, but nonetheless satisfying answers do exist. I believe this chapter has demonstrated that. Getting those answers may take some digging, however. My advice: Never be afraid to dig.

Skeptics have the right to challenge the claims of the Bible. We live in a nation in which people have the right to choose their form of worship, or lack thereof, and to make their voice heard regarding their religious or philosophical worldview. Additionally, and far more importantly, God grants us the free will to choose whether or not we will worship him, or worship a false god or an idol in his place. As Christians we are commanded to "go and make disciples of all nations"[355] and "Go into all the world and preach the good news to all creation,"[356] and we should do so knowing that we will be challenged from many different viewpoints. As Christian evangelists – which, by the way, is all Christians – we must be ready with answers to biblical difficulties, but we must

---

[353] Genesis 6:6, NIV.

[354] Rhodes, 248.

[355] Matthew 28:19, NIV.

[356] Mark 16:15, NIV.

do so with "gentleness and respect."[357]

In the next chapter we will examine the skeptical claim that Christians automatically condemn to Hell those unfortunate souls who never had the chance to hear the Gospel message while still alive on the earth. This chapter will focus on the issue of eternal justice – or, in the eyes of the skeptical, eternal injustice. This has been an issue that I have wrestled with for a long time, and I sincerely hope that you wrestle with it, too. Doing so will make us better-equipped apologists.

---

[357] 1 Peter 3:15, NIV.

CHAPTER EIGHT

One Tough Issue:
The Destiny of the Unevangelized

Among skeptics and believers alike, a particularly troubling question involves the eternal fate of those who never even heard about Christ while alive on the earth. The question of the eternal fate of the unevangelized has been asked many times, and rightfully so, for it is a challenge to our sense of justice. Like the problem of suffering and evil, many skeptics are quick to criticize the biblical faith because of this issue. As defenders of the faith, believers must be ready to tackle this issue head-on – and always be ready with an answer (1 Peter 3:15).

I remember the first time I was confronted with the question of the unevangelized. A woman in one of the classes that I was teaching on world religions confidently told me that the 'heathen' who never had the benefit of hearing the Gospel message while alive on the earth were simply 'out of luck' at their time of death. But how could God allow that kind of situation? Most of us have, at some time or another, joked that we have had bad luck, but bad luck like this is on a cosmic scale of seriousness. Bad luck which lasts for eternity is certainly not to be taken lightly. Up to that time, I had been so preoccupied with learning the background histories and beliefs of the major world religions that I had never taken the time to really consider this question. That was the first time that I really began to get a feel for the burden that is on the heart and mind of every sincere missionary. Since that encounter, I have had more than a few similar experiences which have left me deeply saddened, due to the seeming ease with which some believers assign to Hell untold millions – a task which is not for them to undertake, by the way. I can only hope that these people have taken the time to seriously think through this issue.

Years later I am still exploring this skeptical objection to the faith. There are several different positions that exist within the Christian community, and surprisingly there is a wide range of views even within the evangelical wing of the faith.

Fortunately, many Christians are uncomfortable with automatically condemning to Hell the untold millions who never had the opportunity to hear about Christ through no fault of their own. This issue generates many questions: What about those who lived before Christ? What about the Old Testament 'heroes of the faith,' such as Job, who almost certainly did not know the name of Jesus but nonetheless responded positively to the one true God of the universe? What about those people who lived during the life of Jesus, or even after, but in a place far removed from the Holy Land, in a country where the Gospel had not yet made it during their lifetime? The list of questions is long indeed.

The Old Testament describes the great faith that many non-Israelite men and women possessed, yet they lived long before Christ. They could not possibly be expected to acknowledge the person of Jesus as their Lord and Savior, as he had not yet come into the world. How could they be saved? Many examples of these Old Testament faithful come to mind. Melchizedek is perhaps the first such person that we encounter in Scripture:

> Then Melchizedek king of Salem brought out bread and wine. He was priest of God Most High, and he blessed Abram, saying, "Blessed be Abram by God Most High, Creator of heaven and earth. And praise be to God Most High, who delivered your enemies into your hand." Then Abram gave him a tenth of everything.[358]

Another such person is Jethro: "Then Jethro, Moses' father-in-law, brought a burnt offering and other sacrifices to God, and Aaron came with all the elders of Israel to eat a meal with Moses' father-in-law in the presence of God."[359]

Nebuchadnezzar also comes to mind. He was a man who learned the hard way that the God of Israel is the one true God:

> At the end of that time, I, Nebuchadnezzar, raised my eyes toward heaven, and my sanity was restored. Then I praised the Most High; I honored and glorified him who lives forever. His dominion is an eternal dominion; his kingdom endures from generation to generation. All the peoples of the earth are regarded as nothing. He does as he pleases with the powers of heaven and the peoples of the earth. No one can hold back his hand or say to him: "What have you done?" At the same time that my sanity was restored, my honor and splendor were returned to me for the glory of my kingdom. My advisers and nobles sought me out, and I was restored to my throne and became even greater than before. Now I, Nebuchadnezzar, praise and exalt and glorify the King of heaven, because everything he does is right and all his ways are just. And those who walk in pride he is able to humble.[360]

Naaman comes to mind as well. Like Nebuchadnezzar and the others, he eventually came to realize that the God of Israel is the one true God (2 Kings 5:15).

As mentioned, there are several views regarding the eternal fate of the unevangelized. Perhaps the most popular position is the agnostic view, which maintains that no one really knows the eternal fate of the unevangelized. Those who are theologically agnostic ('without knowledge') on this issue believe that

---

[358] Genesis 14:18-20, NIV.
[359] Exodus 18:12, NIV.
[360] Daniel 4:34-37, NIV.

Scripture is too vague to render a definitive position. Some people are agnostic-pessimistic, believing that people will never be able to arrive at a definitive view on the matter, while others are agnostic-optimistic and confident that, with enough study and insight, someday we all will be able to arrive at a conclusive position. This view likely contains the greatest number of adherents among Christians as a whole, although not necessarily among academically trained theologians and clergy who tend to arrive at a more definitive position on the matter. That 'more definitive position' might be exclusivism, inclusivism, or maybe even one of the other positions that will be described shortly.

The Two Major Views: Exclusivism & Inclusivism

Concerning the eternal fate of the unevangelized, the two big views that are in opposition to each other are exclusivism and inclusivism. Exclusivism is the belief that salvation is attained only by professing the name of Jesus Christ as Lord and Savior, while still alive on the earth. This is the position that skeptics and non-Christians of all varieties find insulting, and it is certainly not the 'politically correct' position in our society today. Award-winning author and apologetics professor Lee Strobel drives' home this point: "Many people consider it arrogant, narrow-minded, and bigoted for Christians to contend that the only path to God must go through Jesus of Nazareth. In a day of religious pluralism and tolerance, this exclusivity claim is politically incorrect, a verbal slap in the face of other belief systems."[361] Even among exclusivists there are variable degrees of hope for the lost, but the general belief is that those who never heard about Christ while they were alive – through no fault of their own, mind you – are doomed to eternal separation from God. Prominent proponents of exclusivism throughout history include Augustine, John Calvin, and R.C. Sproul.

Keep in mind that exclusivism may be the correct position regarding the eternal fate of the unevangelized, regardless of how you personally feel about it. Although many Christians struggle with the idea of God allowing those who have never heard of Jesus to become separated from him forever, we must also remind ourselves that there is much that we do not know about spiritual matters on this side of eternity.

The other major view is inclusivism. Inclusivists maintain that God is more interested in the direction a person was going in their spiritual journey, rather than their spiritual location at the time of death. In other words, how did the person respond to God using the knowledge of God that he or she had at their time of death? Examples of inclusivists throughout history include Irenaeus, Clement of Alexandria, John Wesley, and C.S. Lewis.

Although exclusivists will tell you that inclusivism is scripturally bankrupt, I beg to differ. In fact, I would say that inclusivism is the correct biblical position on this matter. It must be stressed that inclusivists firmly believe that salvation is found in Christ alone: "Salvation is found in no one else, for there is no other

---

[361] Strobel, 146.

name under heaven given to mankind by which we must be saved."[362] Jesus made it clear to us that he alone grants us access to the Father in Heaven: "I am the way, and the truth, and the life. No one comes to the Father except through me."[363]

Besides providing us with the clear means of salvation through Christ alone, it is God's wish for all to be saved:

> I urge, then, first of all, that petitions, prayers, intercession and thanksgiving be made for all people – for kings and all those in authority, that we may live peaceful and quiet lives in all godliness and holiness. This is good, and pleases God our Savior, who wants all people to be saved and to come to knowledge of the truth.[364]

Inclusivists also acknowledge that God is just: "God is just: He will pay back trouble to those who trouble you"[365] and "Great and marvelous are your deeds, Lord God Almighty. Just and true are your ways, King of the nations."[366] Since God wishes for all to be saved through Christ, and since God is just, we should expect him to be both concerned about those who have never had the opportunity to respond to salvation in Christ, and to deal with them fairly.

However, in their fallen (unregenerated) state people are unable to reach out to God on their own accord, in full submission to the Lord: "The wrath of God is being revealed from heaven against all the godlessness and wickedness of people, who suppress the truth by their wickedness."[367] Romans 7:14-15, Romans 8:5-8, and 2 Corinthians 4:4 are other key verses in this regard.

Even without knowing about Christ and his act of redemption, all people do know God through general revelation, which consists of both creation and an inborn knowledge of God. The psalmist wrote that "the heavens declare the glory of God; the skies proclaim the work of his hands,"[368] while Paul's discourse on general revelation may be even more powerful: "For since the creation of the world God's invisible qualities – his eternal power and divine nature – have been clearly seen, being understood from what has been made, so that people are without excuse."[369] Solomon told us that everyone who has ever lived inherently knows the one true God who dwells in eternity: "He has made everything beautiful in its time. He has also set eternity in the human heart; yet no one can

---

[362] Acts 4:12, NIV.
[363] John 14:6, ESV.
[364] 1 Timothy 2:1-4, NIV.
[365] 2 Thessalonians 1:6, NIV.
[366] Revelation 15:3, NIV.
[367] Romans 1:18, NIV.
[368] Psalm 19:1, NIV.
[369] Romans 1:20, NIV.

fathom what God has done from beginning to end."[370] Finally, Paul discusses how God created us with the moral law hardwired into us:

> (Indeed, when Gentiles, who do not have the law, do by nature things required by the law, they are a law for themselves, even though they do not have the law. They show that the requirements of the law are written on their hearts, their consciences also bearing witness, and their thoughts sometimes accusing them and at other times even defending them.)[371]

The skeptic truly has no excuse for denying God's existence. Yet so many of them have somehow convinced themselves that God is nothing more than a fictitious character created in the mind of man.

It seems logical, at least to inclusivists, that God will judge those who never heard the name of Christ based upon what they did with their knowledge of God. Whether you agree with inclusivism or not, this view does confirm for us that not all Christians automatically condemn the unevangelized to eternity in Hell. That is the key point to be made in this chapter.

An interesting version of inclusivism comes from Justin Martyr, who was one of the earliest post-apostolic defenders of the faith. Justin's *prisca theologia* ('ancient theology') in part maintains that God uses the numerous religions and philosophies of the world to prepare people for the Gospel message, although only Christianity wholly fulfills God's plan of salvation. In other words, non-Christian beliefs are merely a 'pointer' to the truth that is found in its entirety in Christ. Justin noted that, although there may be several commonalities between the religions, there are also critical errors between them as well. Therefore, he was not a pluralist who believed that all religions equally lead to God, although he has often been accused of having held to that doctrine.

Justin wrote about the *logos spermatikos* ('seeds of the word') in chapter thirteen of his *Second Apology*. Although lengthy, it is worth quoting the entire chapter here, in three sections:

> For I myself, when I discovered the wicked disguise which the evil spirits had thrown around the divine doctrines of the Christians, to turn aside others from joining them [2 Corinthians 4:4], laughed both at those who framed these falsehoods, and at the disguise itself and at popular opinion and I confess that I both boast and with all my strength strive to be found a Christian; not because the teachings of Plato are different from those of Christ,

---

[370] Ecclesiastes 3:11, NIV.
[371] Romans 2:14-15, NIV.

but because they are not in all respects similar, as neither are those of the others, Stoics, and poets, and historians.[372]

The teachings of Plato are not that different from those of Christ? That should capture your attention! Justin does admit that there are some differences between the two teachers, but in his mind the similarities were due to the *logos*. The *logos* is the Greek term for 'principle of divine reason and creative order inherent in the universe.' As discussed in chapter two, the Apostle John used this Greek term to refer to Christ, to illustrate the point that Jesus is the creator and sustainer of everything. Christ is the *Logos* (Word), but everyone who has ever spoken truth has done so because Christ – who is truth embodied – has made it possible. That is why even the so-called 'barbarian' could figure out theological truths from time to time. The barbarian is, according to Justin, being prepared for the truth of the Gospel in his own way. Justin continues:

> For each man spoke well in proportion to the share he had of the spermatic word [*logos spermatikos*], seeing what was related to it. But they who contradict themselves on the more important points appear not to have possessed the heavenly wisdom, and the knowledge which cannot be spoken against. Whatever things were rightly said among all men, are the property of us Christians.[373]

Justin declared that all truth is God's truth, and therefore all truth is the property of God's people – regardless of who it is spoken by. Even the pagans locked onto truth now and again, which according to Justin prepared them for the greater truths found in the Gospel message. Did all pagans discover 'ultimate truth' at some point? Hardly. But according to Justin, some of them were pointed in the right direction, toward the one true God of the universe, through the limited amount of truth that they had at their disposal. Justin concludes the chapter with these words:

> For next to God, we worship and love the Word who is from the unbegotten and ineffable God, since also He became man for our sakes, that becoming a partaker of our sufferings, He might also bring us healing. For all the writers were able to see realities darkly through the sowing of the implanted word that was in them. For the seed and imitation impacted according to capacity is one thing, and quite another is the thing itself, of which there is

---

[372] Justin Martyr, *Second Apology* (Chapter 13). https://newadvent.org/fathers/0127.htm
[373] Ibid.

the participation and imitation according to the grace which is from Him.[374]

Justin continued to develop the Christian idea of Jesus as the *Logos* (Word), which was begun by John in the previous century (John 1:1; 1 John 1:1). Justin noted that each person spoke truth in proportion to the share he or she had of the *logos*. Justin believed that one did not have to necessarily pass a 'theological test' to enter Heaven, but rather had to do the best he or she could with the amount of spiritual truth they had at their time of death. Regardless of the form that inclusivism may take, the big idea behind this view is this: God is more concerned with where a person was headed spiritually, rather than the persons 'spiritual location' at their time of death.

Some 'Interesting but Scripturally Questionable' Positions

Besides agnosticism, exclusivism, and inclusivism there are other positions concerning the eternal fate of the unevangelized. Each of these has varying degrees of scriptural support. Two views that have captured the interest of believers throughout the centuries are post-mortem salvation and accessibilism. Although both views have some scriptural support, they contain scriptural difficulties at the same time.

There are two closely related positions that are best described as 'post-mortem salvation.' Both views are generally held by conservative theologians to be speculative in nature, but it must be pointed out that the proponents of these views do appeal to Scripture:

> For as Jonah was three days and three nights in the belly of a huge fish, so the Son of Man will be three days and three nights in the heart of the earth...Very truly I tell you, a time is coming and has now come when the dead will hear the voice of the Son of God and those who hear will live...This is why it says: "When he ascended on high, he took many captives and gave gifts to his people." (What does "he ascended" mean except that he also descended to the lower, earthly regions? He who descended is the very one who ascended higher than all the heavens, in order to fill the whole universe.)[375]
>
> For Christ also suffered once for sins, the righteous for the unrighteous, to bring you to God. He was put to death in the body but made alive in the Spirit. After being made alive, he went and made proclamation to the imprisoned spirits – to those who were disobedient long ago when God waited patiently in the days of Noah while the ark was being built. In it only a few people,

---

[374] Ibid.
[375] Matthew 12:40, NIV; John 5:25, NIV; Ephesians 4:8-10, NIV.

eight in all, were saved through water, and this water symbolizes baptism that now saves you also – not the removal of dirt from the body but the pledge of a clear conscience toward God. It saves you by the resurrection of Jesus Christ, who has gone into heaven and is at God's right hand – with angels, authorities and powers in submission to him.[376]

For this is the reason the gospel was preached even to those who are now dead, so that they might be judged according to human standards in regard to the body, but live according to God in regard to the spirit.[377]

The 'final option' position maintains that all people will have an encounter with Jesus Christ at the interface between life and death, and therefore have an opportunity to believe in Christ as their Lord and Savior. Proponents claim three points in support of their view. First, every single person – even those who already confess Christ as their Lord and Savior – will hear about Jesus' work on the cross, from none other than Christ himself. Second, this encounter with Christ occurs at the exact moment when the soul is separated from the body and, for the first time, a person can make a totally free decision without any constraints. Third is the idea that prior spiritual choices may influence, but do not wholly determine, this final decision. If post-mortem salvation is true, it is hard to imagine someone looking the one true God of the universe in the eye and declining his offer of salvation. But, then again, free will in a fallen creature can lead to some strange situations.

The strongest evidence for this position comes from the rather controversial study of near-death experiences.[378] In near death experiences, 'space-time restrictions' seem to break down, if not disappear altogether. I can vouch for this, from personal experience. I nearly drowned as a young boy. Although under the water for only seconds, already having taken in what I considered to be a fair amount of water, I saw my life 'flash before my eyes.' I saw images of my life, going as far back as I could remember. It seemed like this mental video lasted for several minutes, yet I was assured by numerous people that I was underwater for only a short time, not even one minute. In this situation, time was dramatically different from what I knew it to be normally. Regardless of whether post-mortem salvation has any merit, I can easily see how someone in the beginning phase of death – experiencing the time/eternity interface – would have the opportunity to

---

[376] 1 Peter 3:18-22, NIV.

[377] 1 Peter 4:6, NIV.

[378] The near-death experience is controversial because it is a 'near' death experience. In other words, it is not a 'died-and-came-back-from-irreversible-death' experience. Therefore, many researchers hold that these are a meaningless phenomenon, providing no real data on what lies beyond death.

interact with Christ before final death occurs.[379]

The 'second chance' position closely resembles the final option view, except that in the second chance position the opportunity to decide for or against Christ happens sometime after final (irreversible) death, as opposed to the life/death interface. The difference between the two positions lies in the 'timing,' for lack of a better word. I personally prefer the 'final option' view over the 'second chance' position, because scripturally it makes sense that our decision for or against Christ must be made prior to final (irreversible) death.

An important question related to post-mortem salvation is this: Will God redeem those souls already in Hell? Some of the Early Church Fathers believed that some lost souls in Hell could be redeemed from their torment. This is a position with striking similarities to the Roman Catholic doctrine of purgatory, to be addressed shortly. It is held by some theologians today that Clement of Alexandria, Origen, and Athanasius strongly considered this possibility, although this idea faces powerful opposition from Scripture and is labeled heretical by many believers today. Personally, I would not rule it out completely. There is much that we do not know regarding salvation among the unevangelized.

Although considered by many to have been an inclusivist, Irenaeus is also believed to have strongly held to the post-mortem position: "Irenaeus and other early interpreters saw in these passages [1 Peter 3:18-22; 4:6] the proclamation of the gospel to those who had not heard it during their time on earth."[380] Besides the early Church leaders already mentioned, other ancient theologians associated with this position include Melito, Hippolytus, Gregory of Nazianzus, and John of Damascus. After having fallen out of favor in the fifth century, salvation after death was revitalized in the nineteenth century and is currently enjoying a rebirth of sorts among some theologians. This position is also popular today with some Roman Catholic theologians.

The other 'suspect' position in this section is accessibilism, which is most often defined as the view in which those who would respond to God in any possible world are given the opportunity to do so in this one. Stated differently, those who would freely accept Christ in any situation – regardless of the time, place, or culture into which they were born – would be placed in a position to hear about Christ and therefore respond positively. On the flip side, those who would never accept the Gospel message in any situation are assigned by God to a time and place where they will never be exposed to it. (For instance, in the heart of the Assyrian culture centuries before Christ or even before Jonah, who was the great evangelist to the Assyrians in Old Testament times.) These non-Christian, and eternally damned, individuals suffer from what accessibilists have

---

[379] Death is not easily defined in medical terms. Several physiological factors are considered by medical personnel when attempting to confirm physical death, but even these factors are not always well understood scientifically.

[380] Gerald McDermott, *God's Rivals* (Downers Grove, IL: InterVarsity Press, 2007), 111.

termed 'trans-world depravity,' meaning that they would never accept God's saving grace in any situation.

Despite the increasing popularity of accessibilism, there is a point that needs to be considered. As mentioned, there are many non-Jews in the Old Testament who responded positively to the one true God. These men and women lived centuries before Christ, in lands that were ruled by false gods. People such as Job, Melchizedek, Jethro, Nebuchadnezzar, Naaman, and Rahab were separated from Christ by centuries, placed in lands that were polytheistic and generally knew nothing of the one true God. This also applies to the 'righteous pagan' Cicero, who missed Christ by only decades. He did more for the intelligent design movement in pre-Christian Rome than anyone. Yet, despite living decades if not centuries before Christ – and in pagan cultures, at that – these non-Jews responded positively to the Creator of the universe.[381] It seems to me that, had they somehow been made aware of Christ, they would likely have responded to him in a positive manner. Yet many accessibilists maintain that anyone who would reject the one true God of the universe would be placed in a situation where he or she would not be able to respond positively to God. On the flip side, people in the West today must be willing to accept Jesus as Lord and Savior, since they are fortunate enough to live in a time, place, and culture that acknowledges Christ. Yet so many people in America today, despite having an abundance of opportunities to hear about Christ's gift of salvation, choose to decline it anyway.

Accessibilism is also known as Molinism, as this position was initially proposed by the sixteenth-century Spanish Jesuit Luis de Molina. The Christian philosopher William Lane Craig is widely considered to be the preeminent Molinist today. He has written extensively on divine foreknowledge and human free will in books such as *The Only Wise God: The Compatibility of Divine Foreknowledge & Human Freedom.*

Whether you agree or disagree with post-mortem salvation and accessibilism, these views further confirm for us that not all Christians automatically condemn the unevangelized to eternity in Hell. Maybe these views are brand new ideas for you. If so, I hope you investigate them further.

Some Scripturally Untenable Positions

There are some lesser-accepted views among today's evangelicals as well. They are not widely accepted because they are wrought with difficulties. Nonetheless, believers should be aware of them. Included in this discussion is universalism, pluralism, purgatory, and even reincarnation.

Universalism states that all souls will be saved regardless of one's spiritual beliefs while alive. Adolph Hitler, for example, will spend eternity with God – whether he wants to or not. Proponents of universalism rely upon several verses

---

[381] Cicero, as the only non-biblical person in this list, is debatable. However, I personally would not count him out.

(Matthew 18:14; Luke 3:6; John 3:17; 12:32, 47; Romans 5:18; 1 Corinthians 15:22-28; 1 Timothy 4:10; 1 John 2:2; Revelation 21:25).

Of course, opponents of universalism are quick to point out that these verses are lifted out of the greater context of the Bible as a whole. Although it is an encouraging belief for many, universalism does not reconcile with Matthew 25:31-46, which is the parable of the sheep and the goats. That parable ends with these words: "Then they will go away to eternal punishment, but the righteous to eternal life."[382] Likewise, Mark's Gospel account speaks of a sin which is unforgivable: "Truly I tell you, people can be forgiven all their sins and every slander they utter, but whoever blasphemes against the Holy Spirit will never be forgiven; they are guilty of an eternal sin."[383] The doctrine of universal salvation finds its greatest strength in the love of Christ, but Jesus nonetheless spent more time discussing the reality of Hell than he spent proclaiming the awesomeness of Heaven, a point which must not be forgotten.

Origen is the Early Church Father commonly associated with universalism, although Gregory of Nyssa, Theodore of Mopsuestia, and Diodore of Tarsus have been linked to universal salvation as well. In general, the Eastern Church was more inclined towards universalism, since they tended to emphasize God's healing love in salvation, whereas the Western Church viewed salvation more in terms of God's 'legal' justice. From the time of Augustine, the doctrine of universal salvation declined tremendously in the West and was not seriously reconsidered until after the Protestant Reformation, when some Anabaptists began to promote it. Widespread support for universalism began in Protestant theology primarily through the work of Friedrich Schleiermacher, the first post-Enlightenment theologian to push this belief. Karl Barth, William Barclay, Karl Rahner, and Hans Kung are some of the more prominent theologians associated with universalism in the twentieth century. In general, many within the Roman Catholic Church, especially post-Vatican II, have strongly considered this position. I sincerely wish it were true, but this view calls into question the validity of human free will and it flies in the face of many passages of Scripture.

Pluralism, which closely resembles universalism, states that all who faithfully and sincerely obey the tenets of their religion or worldview will be united to God after this earthly life is over. The pluralist will refer to several verses in support of his or her position. Not surprisingly, pluralists utilize many of the same verses that universalists reference, such as these verses from Luke's Gospel account: "Be merciful, just as your Father is merciful. Do not judge, and you will not be judged. Do not condemn, and you will not be condemned. Forgive, and you will be forgiven."[384]

The pluralist, however, must struggle against the fact that many of the world religions hold drastically different views on God. Christianity teaches triune

---

[382] Matthew 25:46, NIV.

[383] Mark 3:28-29, NIV.

[384] Luke 6:36-37, NIV.

monotheism, or one God in three persons. Both Judaism and Islam teach strict monotheism, or one God in one person. Hinduism generally holds to a combination of millions of gods (polytheism), the belief that everything is one (monism), and the belief that everything is divine (pantheism). Buddhism, on the other hand, is functionally atheistic. How can everyone be united to God after this life is over when so many radically diverse views of God exist within the world's religions? John 14:6 is the major stumbling block for pluralists: "Jesus answered, "I am the way and the truth and the life. No one comes to the Father except through me.""[385]

John Hick (1922-2012) is the theologian commonly associated with pluralism in the modern era. Hick is interesting in that he began as a fundamentalist, moved into conservative evangelicalism, and finally settled into full-blown pluralism. Unfortunately, Hick left behind even the most basic doctrines of the Christian faith over time. Other modern-day pluralists include David Strauss, Alexander Pope, and Paul Knitter. From a 'politically-correct' perspective – which carries no weight whatsoever in God's kingdom – it is fair to say that religious pluralism is commonly adhered to by liberal theologians and 'crowd-pleasing' clergy only.

Purgatory is the Roman Catholic belief that people go to an intermediate state between death and Heaven, where one is both punished and purified for transgressions during life on the earth. Purgatory is based upon the apocryphal writing 2 Maccabees 12:39-45:

> On the next day, as by that time it had become necessary, Judas and his men went to take up the bodies of the fallen and to bring them back to lie with their kinsmen in the sepulchres of their fathers. Then under the tunic of every one of the dead they found sacred tokens of the idols of Jamnia, which the law forbids the Jews to wear. And it became clear to all that this was why these men had fallen. So they all blessed the ways of the Lord, the righteous Judge, who reveals the things that are hidden; and they turned to prayer, beseeching that the sin which had been committed might be wholly blotted out. And the noble Judas exhorted the people to keep themselves free from sin, for they had seen with their own eyes what had happened because of the sin of those who had fallen. He also took up a collection, man by man, to the amount of two thousand drachmas of silver, and sent it to Jerusalem to provide for a sin offering. In doing this he acted very well and honorably, taking account of the resurrection. For if he were not expecting that those who had fallen would rise again, it would have been superfluous and foolish to pray for the dead. But if he was looking to the splendid reward that is laid up for those who fall asleep in godliness, it was a holy and pious thought.

---

[385] John 14:6, NIV.

Therefore he made atonement for the dead, that they might be delivered from their sin.

According to this belief, after a certain time – which varies depending upon how good or bad one was during life – all sin is atoned for, and then one is worthy to enter into God's presence. The souls in purgatory are helped along in their pursuit of purification by the prayers of those who are still alive on the earth.

The sad fact of the matter is that purgatory attempts to negate the substitutionary atonement of Christ. Christ paid for our sins in full on the cross, therefore purgatory is unnecessary:

> But if we walk in the light, as he is in the light, we have fellowship with one another, and the blood of Jesus, his Son, purifies us from all sin.[386]
>
> For it is by grace you have been saved, through faith – and this is not from yourselves, it is the gift of God – not by works, so that no one can boast.[387]
>
> But when the kindness and love of God our Savior appeared, he saved us, not because of righteous things we had done, but because of his mercy. He saved us through the washing of rebirth and renewal by the Holy Spirit, whom he poured out on us generously through Jesus Christ our Savior, so that, having been justified by his grace, we might become heirs having the hope of eternal life.[388]

Some may be inclined to associate purgatory with the post-mortem views, especially the second chance position. The difference between these positions, however, is that in the second chance view non-Christians are given the opportunity for salvation which will lead immediately to eternal communion with God, whereas in purgatory one must first be purified before he or she is worthy of entering into the presence of God – which requires 'time,' for lack of a better term as physical time will not exist in eternity. (Time, space, and matter-energy are the three components of the physical universe.) Even between these two views, it still comes down to grace (second chance) versus works (purgatory).

Amazingly, some professing Christians have attempted to combine their biblical beliefs with the Eastern concept of reincarnation. Reincarnation, also known as transmigration of the soul, is the belief that through a series of deaths and rebirths, a soul will eventually purge itself of all of its sins ('negative karma'), which allows it to finally be united to the 'Ultimate Reality,' which is

---

[386] 1 John 1:7, NIV.
[387] Ephesians 2:8-9, NIV.
[388] Titus 3:4-7, NIV.

nothing more than the monistic-pantheistic god-force. For many adherents of reincarnation, upon uniting with the Ultimate Reality the individual soul loses its distinctly unique personality and becomes one with this cosmic force. In terms of practicality, one may as well subscribe to annihilation of the soul, to be described shortly, since the individually unique spirit that constitutes the real person ceases to exist anyway. Hebrews 9:27-28 is a clear refutation of reincarnation: "Just as people are destined to die once, and after that to face judgment, so Christ was sacrificed once to take away the sins of many; and he will appear a second time, not to bear sin, but to bring salvation to those who are waiting for him."[389]

Only the Eastern-influenced, pseudo-Christian groups such as Unity School of Christianity show an interest in blending reincarnation with Christian doctrine. Most people within Christianity who hold to this idea are self-styled 'New Age Christians' who erroneously believe that one can successfully blend the tenets of the Eastern religions with Judeo-Christian beliefs. No serious evangelical scholars hold to this position.

The views in this section have serious theological problems, but once again these views further confirm for us that not all those who profess Christ as their Lord and Savior automatically condemn the unevangelized to eternity in Hell. Although I cannot speak positively about pluralism, purgatory, or reincarnation, I really do find myself drawn to universalism. I do not accept it, mind you, but I do find myself drawn to it. I have a few friends who are universalists, and I understand why they believe in this doctrine: It is comforting, and it completely squares with the loving nature of God. They have more than a few verses in their favor as well. However, the entirety of Scripture, combined with our free will, insists that Hell is real. But what is really meant by 'Hell,' anyway? We will examine this question in a moment. But first, let us look at how God makes himself known to the seeker of truth.

God Will Make Himself Known

There is one more view concerning the eternal fate of the unevangelized, and it is scripturally solid. The universal presentation view teaches that God will see to it that all non-Christians who seek after the one true God will be exposed to the Gospel message before they die. (At least, the Gospel message in some way, shape, or form that is understandable to the seeker, given the amount of spiritual knowledge that he or she has at their disposal. This would account for their time, place, and cultural understanding of God and salvation.) According to proponents of this view, no one is eternally damned without the opportunity of being saved in this life. While they are still alive on the earth, these non-Christians must not only hear but believe the Gospel message to be saved. Some who hold this view believe that God sends the Gospel only through human

---

[389] Hebrews 9:27-28, NIV.

messengers, while others believe that God may use angels, dreams, or visions.[390] I believe that God can use whatever means he so desires to impart the message of Christ to the seeker of truth. We should never attempt to limit God in any way.

Universal presentation differs from inclusivism in that universal presentation requires that the lost must hear about, and respond positively to, the Gospel message while alive on the earth. Explicit knowledge of Christ is necessary for salvation in this view. Like exclusivism, universal presentation insists that human destinies are sealed at death, so non-Christians must hear about, and respond in faith to, the Gospel message prior to death. This view denies post-mortem salvation, of course. If anyone truly seeks God, God will ensure that the Gospel message will be given to that person while he or she is still alive.

Unlike the strict, seemingly unbending position of exclusivism, universal presentation truly does appeal to God's love and justice. Many of the adherents of this view are quick to cite the story of Philip and the Ethiopian eunuch in support of their position:

> Now an angel of the Lord said to Philip, "Go south to the road – the desert road – that goes down from Jerusalem to Gaza." So he started out, and on his way he met an Ethiopian eunuch, an important official in charge of all the treasury of the Kandake (which means "queen of the Ethiopians"). This man had gone to Jerusalem to worship, and on his way home was sitting in his chariot reading the Book of Isaiah the prophet. The Spirit told Philip, "Go to that chariot and stay near it." Then Philip ran up to the chariot and heard the man reading Isaiah the prophet. "Do you understand what you are reading?" Philip asked. "How can I," he said, "unless someone explains it to me?" So he invited Philip to come up and sit with him…Then Philip began with that very passage of Scripture and told him the good news about Jesus.[391]

Likewise, proponents of this view also appeal to the story of Peter's encounter with the Roman centurion Cornelius, who was blessed by Peter's presentation of the Gospel message (Acts 10). Universal presentation also finds support in the books of Jeremiah and Revelation:

> You will seek me and find me when you seek me with all your heart.[392]

---

[390] For instance, many former Muslims who came to Christ claim that they were given the message of the Gospel in a vision or a dream, by none other than Christ himself.
[391] Acts 8:26-31, 35, NIV.
[392] Jeremiah 29:13, NIV.

> Here I am! I stand at the door and knock. If anyone hears my voice and opens the door, I will come in and eat with that person, and they with me.[393]

Christ is proclaiming that the sincere seeker of truth, open-minded and willing to go wherever led by God, will find the Lord because he will see to it that the seeker has that opportunity. Therefore, those who hold firmly to universal presentation are adamant that no one will be separated from God for eternity who should not be, since everyone who has ever lived has had the ability to enter into a relationship with the one true God if they so desired.

The Medieval theologian Alexander of Hales proclaimed the central tenet of this view when he wrote, "If he [the true seeker] does what is within his power, the Lord will enlighten him with a secret inspiration, by means of an angel or a man."[394] Universal accessibility to the Gospel message became popular in the twelfth through fifteenth centuries and was the position of Peter Lombard, Albert the Great, Saint Bonaventure, Duns Scotus, and Gabriel Biel among others. Thomas Aquinas, considered by some to have been an inclusivist, also seems to have favored a belief in universal presentation. (Not surprisingly, his teacher Albert the Great is also considered to have favored both positions as well). This view is also associated with Alighieri Dante and James Arminius. The late Norman Geisler outlined this position very well in *If God, Why Evil?*[395]

Table 8.1

Views on the Fate of the Unevangelized

EXCLUSIVISM	ONE MUST PROFESS CHRIST AS SAVIOR WHILE ALIVE ON THE EARTH. SCRIPTURAL SUPPORT: EXTENSIVE. PROPONENTS: AUGUSTINE, JOHN CALVIN, AND R.C. SPROUL.
INCLUSIVISM	GOD GRANTS SALVATION TO THOSE WHO DID NOT KNOW ABOUT CHRIST BASED UPON WHAT THEY DID WITH THE AMOUNT OF REVELATION THEY HAD.

---

[393] Revelation 3:20, NIV.

[394] John Sanders, "Those Who Have Never Heard: A Survey of the Major Positions." https://rsc.byu.edu/archived/salvation-christ-comparative-christian-views/those-who-have-never-heard-survey-major-positions

[395] Norman L. Geisler, *If God, Why Evil?* (Bloomington, MN: Bethany House Publishers, 2011), 115-122.

SCRIPTURAL SUPPORT: EXTENSIVE.

PROPONENTS: JUSTIN MARTYR, JOHN WESLEY, AND C.S. LEWIS.

UNIVERSAL PRESENTATION	GOD WILL SEE TO IT THAT ALL NON-CHRISTIANS WHO SEEK AFTER THE ONE TRUE GOD WILL BE EXPOSED TO THE GOSPEL MESSAGE BEFORE THEY DIE.

SCRIPTURAL SUPPORT: EXTENSIVE.

PROPONENTS: ALEXANDER OF HALES, THOMAS AQUINAS, ALIGHIERI DANTE, JAMES ARMINIUS, AND NORMAN GEISLER.

POST-MORTEM SALVATION	THE UNEVANGELIZED MAY ACCEPT CHRIST AT THE INTERFACE BETWEEN LIFE AND DEATH (FINAL OPTION) OR AT SOME POINT AFTER FINAL (IRREVERSIBLE) DEATH (SECOND CHANCE).

SCRIPTURAL SUPPORT: YES, BUT IT IS LIMITED.

PROPONENTS: MELITO, HIPPOLYTUS, AND JOHN OF DAMASCUS.

ACCESSIBILISM	GOD PLACES THOSE WHO ARE OPEN TO SALVATION IN THE TIME AND PLACE THAT THEY NEED TO BE IN, TO HEAR THE GOSPEL OR RECOGNIZE THE ONE TRUE GOD.

SCRIPTURAL SUPPORT: YES, BUT IT IS LIMITED.

PROPONENTS: LUIS DE MOLINA AND WILLIAM LANE CRAIG.

UNIVERSALISM	EVERYONE WHO HAS EVER LIVED WILL BE SAVED.

SCRIPTURAL SUPPORT: YES, BUT SCRIPTURALLY REFUTED AS WELL.

PROPONENTS: ORIGEN, FRIEDRICH SCHLEIERMACHER, KARL BARTH, KARL RAHNER, AND HANS KUNG. MANY ROMAN CATHOLICS, ESPECIALLY POST-VATICAN II, ENTERTAIN THE IDEA OF UNIVERSAL SALVATION.

PLURALISM
EVERYONE WHO SINCERELY FOLLOWS THEIR RELIGION OR WORLDVIEW WILL BE SAVED. 'ALL PATHS LEAD TO GOD.'

SCRIPTURAL SUPPORT: SCRIPTURALLY UNSOUND.

PROPONENTS: JOHN HICK, DAVID STRAUSS, AND PAUL KNITTER.

PURGATORY
AFTER LIFE, PEOPLE SPEND TIME IN A 'HOLDING PLACE' UNTIL THEY ARE SPIRITUALLY WORTHY TO ENTER HEAVEN.

SCRIPTURAL SUPPORT: SCRIPTURALLY UNSOUND.

PROPONENTS: ROMAN CATHOLICS.

REINCARNATION
PEOPLE CAN ENTER PARADISE ONLY AFTER THEIR NEGATIVE KARMA – WHICH HAS BEEN GENERATED OVER SEVERAL LIFETIMES – NO LONGER EXISTS.

SCRIPTURAL SUPPORT: SCRIPTURALLY UNSOUND.

PROPONENTS: 'NEW AGE' CHRISTIANS.

Hell: Eternal Punishment or Eternal Consequence?

Almost all people, whether believers or unbelievers, have at some time or another wrestled with the doctrine of Hell. Do the unsaved experience pain and suffering for eternity, or is the punishment for rebellion against God enacted in some other way? Annihilationism, also known as conditional immortality, is the belief that at death God grants immortality to believers while destroying (annihilating) the souls of hardened unbelievers, rather than keeping their souls alive to suffer for eternity. Although everyone survives death and participates in the final judgment, the verdict passed on unbelievers is 'soul extinction.'

Annihilationist's maintain that the rebellious are punished eternally, but they are not enduring eternal punishment. Theologian Greg Boyd, who may be considered one of the preeminent annihilationist's today, states, "that the wicked are punished eternally, but not that the wicked endure eternal punishment. The wicked suffer eternal punishment (Matthew 25:46), eternal judgment (Hebrews 6:2), and eternal destruction (2 Thessalonians 1:9) the same way the elect experience eternal redemption (Hebrews 5:9; 9:12)."[396] Since the elect do not undergo an eternal process of redemption – once they are redeemed it is forever – likewise unbelievers do not undergo an eternal process of punishment. When unrepentant, unbelieving souls are destroyed, the consequence is forever. Boyd stresses that Hell is eternal in consequence, but not in duration. In other words, "the wicked are destroyed forever (Psalm 92:7), but they are not forever being destroyed."[397]

Jesus told his disciples not to fear those who may kill the body but are not able to kill the soul, but rather fear the one who can destroy both the body and the soul in Hell (Matthew 10:28). Adherents of this view are quick to point out that this destruction of the soul implies annihilationism.

Boyd stresses two points in his theological reasoning for annihilationism. First, in the traditional view of Hell the wicked and unrepentant are not being punished to learn something valuable, as there is nothing remedial about their eternal torment. Rather, the only purpose for Hell is to experience pain – and for eternity, at that. "After twenty trillion trillion years of torment, the damned are no closer to completing their dire sentence than they were their first moment of horror."[398] Annihilationist's maintain that this view is not compatible with a God who, out of love, sacrifices himself for the world. I agree. We must ask ourselves, does John 3:16 square with conscience torment for eternity? If someone lived a good life,[399] and his or her only crime was failing to believe in the God of the Bible, is that unbelief worthy of conscience torment forever? Call me a heretic, but I have a lot of trouble with that belief. If a rebellious soul is hopelessly antagonistic to God, it seems likely to annihilationist's that God would put them out of their misery rather than torment them for eternity. It is true that God cannot allow a hardened, unbelieving soul to inhabit Heaven along with the truly faithful, but would God want to see that soul suffer forever? Especially with no purpose to that suffering, other than pure torment. God is not capricious or evil.

Second, how can all things be reconciled to God (Acts 3:21; Colossians 1:20) if there is a place of eternal torment that co-exists forever alongside Heaven? Boyd notes, "If the traditional view of hell is correct, God remains non-

---

[396] Greg Boyd, "The Case for Annihilationism." https://reknew.org/2008/01/the-case-for-annihilationism/
[397] Ibid.
[398] Ibid.
[399] That is, good by purely human standards, as no one is truly 'good' theologically speaking (Romans 3:9-20).

victorious. Instead of a glorious universal Kingdom unblemished by any stain, an ugly dualism reigns throughout eternity."[400] I do not know about you, but this is a point I cannot dismiss.

On the flip side, however, there are three passages of Scripture that present difficulty for annihilationist's: Matthew 25:46, Revelation 14:9-11, and Revelation 20:10. Matthew 25:46 seems to clearly drive home the point that separation from God for unbelievers is eternal: "Then they [unbelievers] will go away to eternal punishment, but the righteous to eternal life."[401] This verse may denote an eternally enduring punishment for the unrepentant soul, but then again maybe not. Annihilationist's insist that eternal punishment is eternal only in consequence, not in duration.

Beyond the scriptural case against annihilationism, opponents of this view point out that if this belief is true, the fear of Hell becomes undermined. However, not all annihilationist's deny that the wicked will suffer prior to being annihilated. However it might work out, God's justice will be served. (We do not fully understand how God's justice will play out, so we should not be too closed-minded about this point.) Additionally, annihilationist's point out that the traditional teaching on Hell often has no effect on unbelievers whatsoever. In fact, the idea of eternal conscience torment seems to produce the opposite effect in unbelievers: The notion of never-ending punishment radically opposes people's ordinary, inherent sense of justice. As a result, many unbelievers do not take seriously the idea of eternal punishment in Hell. Have you ever wondered why so many people fail to believe in Hell? It is not just unbelievers who dismiss the doctrine of Hell, but many professing Christians also doubt or disbelieve in this doctrine as well. The idea that the little old lady across the street, who gave tirelessly of her time and resources to help those in need and lived a life of service to her fellow man, should suffer torment forever only because she failed to acknowledge Christ as her Lord and Savior – even though she believed in a divine Creator of some sort – is an idea that does not work for a lot of us.

Some of you reading this will be infuriated at this point. Trust me, I know. I have been around the block a time or two concerning theological debates. Some Christian believers want Hell to be defined by eternal conscience torment. (A thought which bothers me tremendously. It might be real, but do we really want it to be real?) "But eternal conscience torment is what the Bible teaches," they say. Am I saying that Hell is not real? No. Am I saying that everyone who has ever lived will waltz right into Heaven when this life is over? No. Am I saying beyond a shadow of a doubt that annihilationism is true? No. I strongly suspect it is, but if you take the time to read what I have written in this chapter and the previous one you should see that I am not making these types of definitive claims. What I am saying is this: Take the time to put your 'non-negotiables' about Hell to the side and open your heart and mind to the possibility of annihilationism. If

---

[400] Boyd.
[401] Matthew 25:46, NIV.

anything, maybe you will begin to understand why some people believe this doctrine, even if you persist in rejecting it.

Several of the Early Church Fathers are considered to have at least entertained the idea of conditional immortality. Justin Martyr may have rejected the Platonic view of the soul's immortality, but it is difficult to say for sure as his writings are somewhat vague in this area. Most biblical scholars believe that it was Arnobius of Sicca, in the early fourth century, who offered the first unmistakable defense of this belief:

> For theirs is an intermediate state, as has been learned from Christ's teaching; and they are such that they may on the one hand perish if they have not known God, and on the other be delivered from death if they have given heed to His threats and proffered favors. And to make manifest what is unknown, this is [unrepentant] man's real death, which leaves nothing behind. For that which is seen by the eyes is only a separation of soul from body, not the last end-annihilation.[402]

From the time of the Reformation onward, the traditional view of Hell – that is, eternal conscious torment – has been the commonly-held position among most Christian groups, although some Adventists have promoted the belief in annihilationism. John Wenham outlined annihilationism in his 1974 book *The Goodness of God*, establishing this position within evangelical Christianity in modern times. Besides Wenham and Boyd, two other prominent theologians – John Stott and Clark Pinnock – have also sought widespread acceptance for this view.

In 1999, a group of scholars from the Evangelical Alliance discussed annihilationism, and concluded that this is a valid view, although it remains a minority position among scholars. Despite what any group or council has to say, ultimately the validity of annihilationism – or any other faith position, for that matter – is dependent upon the scriptural-theological case that may be advanced in its favor.

I personally favor annihilationism over eternal conscious torment. Therefore, I can say with certainty that at least one Christian does not automatically condemn the unevangelized to eternity in Hell.[403]

A Review of the Various Positions

As a review of how the various positions might play out, let us look at the hypothetical example of a man from the distant past who could not possibly have

---

[402] Arnobius, *Against the Heathen* (Book 2, Chapter 14).
https://newadvent.org/fathers/06312.htm

[403] Of course, there are many theologians, pastors, and apologists who favor annihilationism over conscious eternal torment.

known about Christ's redemptive work upon the cross. Let us say that this man lived deep within Africa, a century before the time of Christ. He lived his whole life in a land ruled by animism and polytheism, yet he strongly suspected that there is one powerful Spirit behind the creation of everything. He lived much like his ancestors did, participating in the accepted religious rituals of his people, although unlike most of his people he recognized a supreme Great Spirit behind everything. As a result of this conviction, he sought to live a moral life, exhibiting a personal ethic that exceeded those around him. Nonetheless, he never accepted Christ as his Lord and Savior, which would be an impossibility as he lived before Christ in a country that knew nothing of biblical truth. How might his eternal fate play out? According to the various positions described in this chapter, this man may have experienced salvation in one or more of the following ways:

Agnosticism

According to agnosticism, no one can really know what this man's eternal fate is. Agnosticism is the default position of many believers on this issue, and for good reason: Ultimately, no one can ever say for sure what the eternal fate of the unevangelized will be. Only God knows.

Exclusivism

Despite being a good man who was spiritually open to truth, he did not hear the message of the Gospel proclaimed during his earthly life – through no fault of his own, of course – and therefore he became separated from God for eternity.

Inclusivism

God likely ushered this man into his eternal presence, since he responded positively to the general revelation of God that he had at his disposal.

Post-Mortem Salvation

At some point in the afterlife, whether at the interface between life and death or perhaps even past the 'point of no return,' God himself proclaimed the Gospel message to this man. He almost certainly accepted God's gift of salvation in Christ. How could he not, when standing in awe before his Creator?

Accessibilism

Since this man was born into a time and place that could not possibly allow him to hear and understand the Gospel message, it is likely that he would not have responded positively to the message of the cross, even if he had heard it. (Does that seem reasonable in this case?)

Universalism

He was destined for Heaven, regardless of how he lived his earthly life.

Pluralism

Since this man lived in accordance with the traditional worldview of his culture, he is certainly in Heaven.

Purgatory

After doing 'hard time' in an otherworldly holding place, he will eventually become pure and acceptable to God. When that happens, he will be ready to spend the rest of eternity in close relationship to his Creator.

Reincarnation

At death, this man will 'do life all over again' until he finally sheds the last of his negative karma and enters a blissful union with the Ultimate Reality. Of course, when this happens his distinct, earthly personality is extinguished, so he will essentially experience 'personality annihilationism.'

Universal Presentation

Since this man was aware of the 'Great Spirit' behind everything, and sought to live a moral life, he likely would have accepted Christ had he been fortunate enough to have had the opportunity to hear the Gospel message while still alive. Of course, God in his omniscience is thoroughly aware of this, and therefore would see to it that this man received an adequate amount of divine revelation. (Christ's sacrifice would not take place for another century, so God had to give him a proper theological understanding in that 'BC' context.) Since this man sought the Great Spirit's desire for his life, he was given an opportunity to understand – and, more importantly, accept – God's gracious salvation.

Annihilationism

If for any reason God does not consider this man to be redeemable – which would seem unlikely in this case – the man may either experience punishment for a 'time' and then his soul will be destroyed (annihilated) by God, or he will immediately be extinguished by God, with no time spent in punishment.

I suspect that it comes as no surprise to the reader that I have encountered more than a few exclusivists who much too easily assign untold millions to Hell for not having accepted Christ as their Lord and Savior while alive on the earth, even if there is no way possible that they could have known about Jesus. It is true that these encounters have served, in large part, as the impetus for my exploration of this issue. (Their belief, and sometimes their attitudes, have bothered me greatly at times.) However, a few other factors are equally, if not more important, in my pursuit of an apologetic concerning this matter. First, the God that I worship is the one true God of the universe, and he is known for his love and justice – although he is also known as being wrathful at times, and for good reason. I needed to determine for myself, let alone for others, how I could reconcile God's love and justice with the eternal fate of the unevangelized while remaining true to Scripture as a whole. Second, the most challenging apologetic issues are those that wrestle with the question of a good God in a fallen world, and these are the questions that seekers and skeptics alike also wrestle with and want answers to, and they are looking to believers for these answers – even if believers may be confrontational at times. Let me put it this way: Compared to the focus of this chapter, the matter of evolutionism versus creationism is simple and straightforward. Maybe that is why I am so drawn to the issue of origins: It really is easy compared to other theological issues!

Concluding Thoughts

So, do all Christians automatically assign to Hell those unfortunate souls who never had a chance to hear about Christ while alive on the earth? As we have seen, that is most definitely not the case. You may be agnostic on this issue, or you may be an exclusivist, or maybe you are considering the merits of one of the other positions described in this chapter. But the main point is this: Not all Christians are quick to condemn to Hell the unevangelized, despite what some skeptics claim. God's unwavering justice will be brought to completion. Someday, in the eternal life to come, we will understand just how it all worked out regarding the eternal fate of the unevangelized. But if you are reading this, that day has not yet come, and we will therefore continue to wrestle with this issue.

Although we cannot know with absolute certainty how God will deal with the unevangelized, it is reassuring to know that God is just, perfectly wise, and all-knowing, and therefore all people who have ever lived will spend eternity exactly where they should. It is not heretical to believe that, since God has revealed himself through general revelation to all people, then all people who have not heard the message about Christ may have the opportunity to respond to the Creator through general revelation or even some other means. However, it is a different story for those who have clearly heard the message about Christ's redemption and nonetheless choose to remain steadfast in their disbelief.

We will now turn our attention to the evidence for the soul and the afterlife. Although I felt the book was getting longer than I had originally intended, I needed to include the next chapter to make this one complete. (After all, why should a skeptical reader be concerned about the eternal fate of the unevangelized if they are not convinced there is even an afterlife in the first place?) I also include a different twist on the 'Case for Christ' in this upcoming chapter, which is the most important topic any apologist can cover. Without offering the reader the real reason for believing in Christianity, I will have failed in my duty as an evangelist. (Which, by the way, is every professing believer.)

A lot of secular-minded people believe that Heaven is just a myth for the 'unenlightened.' Skeptics may legitimately question the evidence for the immortal soul and life after death, but as you will see there is a powerful case to be advanced in favor of Heaven. I hope this next chapter touches the heart of everyone who reads it.

CHAPTER NINE

The Ultimate Issue: Death and the Afterlife

During the time I wrote an early draft of this book, Ravi Zacharias – considered by many believers to be the best apologist in the world – passed after a short battle with an aggressive cancer. He was a prolific author, and every book he wrote was illuminating. But maybe the best book he ever penned was *Can Man Live Without God?* Of course, the short answer to that question is 'yes.' People do it all the time. Like Zacharias, I do not recommend it, however. Many people choose to live a secular lifestyle, totally devoid of God or anything spiritual – and they miss out on the greatest thing we can experience in this life.

In this chapter, I want to ask a different-but-related question: Can man die without God? Once again, and sadly, the short answer is 'yes.' People do it all the time. And, once again, I do not recommend it. Since all people know that God exists, there is no reason for this situation to happen. Paul knew that if he remained alive on the earth, he had Christ to guide him along in his journey. And when he died, then he would experience Christ face-to-face. It was a win-win situation: "For to me, to live is Christ and to die is gain."[404] We can have the same.

Eternity in our Hearts

Since the beginning of time, people have wondered if life ends at the grave, or continues in some way after death. Although many books of the Bible address this question, Job tackles the issue of life after death better than most. Early in the book the main character Job asks a timeless question: "When a man dies, will he live again?"[405] Throughout history, people have answered that question with either a resounding 'yes' or a seemingly confident 'no.' The atheist view of death was well represented by Epicurus. Three centuries before Christ, he wrote, "Death does not concern us, because as long as we exist death is not here. And once it does come, we no longer exist." For Epicurus and his followers, death is not to be worried about, because we cannot prevent it from eventually happening. (Which is true, of course.) And once death does come, we are gone – completely gone – and therefore we will never know it has come. For the atheist, when a person dies, that is it: The body begins its descent into decay, and the mind immediately ceases to exist. Game over. The 'Grand Finale.' Sayonara. There are probably many more clichés that could be used, but the point is this: Atheism insists that there is no afterlife. Human beings, and in fact all living creatures, are merely biological machines with no immortal or spiritual component that survives death. Since atheism denies the existence of God and everything spiritual, there can be no immortal soul which continues after this life.

---

[404] Philippians 1:21, NIV.
[405] Job 14:14, NIV.

The late Grant Jeffrey, a prolific author who specialized in apologetics and end-time issues, refutes the atheist view through one of the best – and most-enduring – lines of evidence for the existence of the immortal soul:

> Perhaps the strongest evidence supporting the truth of immortality is that virtually every tribe, nation, and culture throughout history has expressed a strong faith in the reality of a life after death. In addition to the almost universal belief in God, the conviction that we will live again after our earthly bodies return to dust is the strongest and most commonly held belief of humanity. For thousands of years the vast majority of people have approached their personal valley of death with the firm expectation that they will ultimately rise from death and live forever in a better world. The longing for eternal life is the strongest instinct found in the heart of every human.[406]

Jeffrey was correct in stating that the longing for eternal life is found within every human heart, for he is simply restating what King Solomon wrote approximately three thousand years earlier: "He has also set eternity in the human heart…" (Ecclesiastes 3:11). God has hard-wired us to know that we will live on after this life is over. King David, the father of Solomon, wrote, "As the deer pants for streams of water, so my soul pants for you, my God."[407] Solomon, writing in Ecclesiastes, was merely echoing what his father already knew before him: We are designed to be in communion, or close relationship, with God. We are made in God's image (Genesis 1:27), and the Lord designed us to know that we will live on after these earthly lives have come to an end (Ecclesiastes 3:11). Therefore, despite our fallen nature we desire to be at peace with the Lord. Have you ever met an angry atheist? I have. I could attempt to guess their motivation, but the bottom line is this: They lack God's peace. They say they do not want that peace, yet deep down they give every indication that they do. Their actions betray their words.

(Actually, this dual desire – the desire to be at peace with the Lord, and the desire to be free from accountability to him – reflects the biblical description of mankind's nature more convincingly than anything else does. We are created in God's image, and therefore we want to be in close relationship with the Lord. At the same time, we are fallen in nature and therefore we rebel against God. Likewise, we inherently know that we were made for another world (Ecclesiastes 3:11; Philippians 3:20), yet we desire to cling to this one – as if nothing else awaits us on the other side of death. Being 'created in God's image' but having a fallen nature makes for a very contradictory life for every one of us. I address

---

[406] Grant Jeffrey, *Journey into Eternity: Search for Immortality* (Toronto, Ontario: Frontier Research Publications, Inc., 2000), 21-22.
[407] Psalm 42:1, NIV.

this issue in more detail in my first book.[408])

Many centuries after David and Solomon, Augustine wrote in *The Confessions*, "You have made us for yourself, and restless is our heart until it comes to rest in you."[409] Unless we are in communion with God, we will not have the peace of mind that we crave as human beings. Blaise Pascal also wrote about this need to be in communion with God:

> What else does this craving, and this helplessness, proclaim but that there was once in man a true happiness, of which all that now remains is the empty print and trace? This he tries in vain to fill with everything around him, seeking in things that are not there the help he cannot find in those that are, though none can help, since this infinite abyss can be filled only with an infinite and immutable object; in other words by God himself.[410]

R.C. Sproul supported Jeffrey's assertion that the strongest support for the truth of immortality is to be found in human history. He pointed out that the greatest thinkers throughout time have struggled over the question of life after death: "From ancient times the keenest minds of mankind have sought intellectual evidence for the survival of the soul or spirit beyond the grave…Scholars have given the question serious attention because it is the most serious of all questions."[411] It is the most serious of all questions, because each of us faces our mortality every day of our lives. But this is the interesting thing about the so-called 'big questions of life': If you fail to correctly answer the question of origins, you will also fail to correctly answer the question of destiny, or what happens to us after death.

This universally inherent knowledge of eternity is one more line of evidence for God, for if Heaven exists then there is both a natural and a supernatural realm of existence. Although one may try to argue that the natural realm could be the result of chance, random processes of nature, the existence of a supernatural realm only makes sense in the light of God's existence.

This line of evidence is inherently built into us. I am convinced that the atheist must actively suppress this inborn evidence for God and the spiritual realm; knowledge of the spiritual is simply part of who we are. Have you ever been to a funeral? Most of us, unfortunately, will answer that question with a 'yes.' Rarely do you hear someone proclaim that the deceased is now 'dead as a doornail.' That would be extremely poor funeral etiquette, of course. Some people might think that way, but most people seem to be convinced that life in some way, shape, or form will continue after death. It is a natural part of who we

---

[408] Hroziencik, 87-89.
[409] Augustine, *The Confessions* (Book 1, Chapter 1). https://newadvent.org/fathers/110101.htm
[410] Blaise Pascal, *Pensees* (New York, NY; Penguin Books, 1966), 75.
[411] R.C. Sproul, *Reason to Believe* (Grand Rapids, MI: Zondervan, 1978), 146.

are as human beings. But why is that? Because God hard-wired us to know two things: One, he exists, and two, so will we after our earthly lives are over (Ecclesiastes 3:11).

The transcendent view, on the other hand, maintains that a realm of existence awaits us beyond these earthly lives. The transcendent view is not just the biblical view, but the view of anyone who believes that this life is not all there is. This is the view of most people who have ever lived. For people who hold this view, there is a spiritual component to our existence which survives death, and most often this spiritual component enters a heavenly kingdom which is unlike anything experienced here on the earth. Paul represents the transcendent view better than anyone ever has: "No eye has seen, no ear has heard, no mind has conceived the things God has prepared for those who love him."[412] There is also the possibility of eternity separated from God, but that is not the focus of this chapter. (We examined the reality and nature of Hell in the previous chapter.)

The issue of life after death is dependent upon the existence of the soul: No soul, no life after death. Like the atheistic and transcendent views of life after death, mankind has two basic views concerning the soul: The view that the soul is nothing more than the thinking brain, which perishes along with the physical body at death, and the view that the soul is our immaterial component that survives death and lives on forever. In the context of our study, I prefer to call these views the 'evolutionism view' and the 'creationism view.' These two views have been in collision for a long time.

When Jesus spoke the words, "What good will it be for someone to gain the whole world, yet forfeit their soul?"[413] he presupposed his hearer's belief in the immortal soul. However, after much religious skepticism in our culture – going back at least to the time of the Enlightenment, and likely even further back than that – many people will need to examine the issue of the soul's existence in some detail.

I contend that the atheist worldview is bankrupt. And if atheism can be refuted, there is every reason to believe that the soul is real. After all, "with God all things are possible."[414]

I have three reasons for believing in the reality of the soul and the afterlife. One, the atheist worldview has been thoroughly refuted. Two, people inherently understand that we were created for eternity. Three, the resurrection of Jesus Christ shows us that life continues beyond the grave. You may have different, or additional, reasons for believing in the afterlife. For example, the evidence from near-death experiences seems to work for a lot of people. (I also find this line of evidence for the soul intriguing.) But for me, my three reasons give me everything I need. I hope that these three reasons for the soul and the afterlife are an encouragement to your faith as well. Or, for those readers who lack faith so

---

[412] 1 Corinthians 2:9, NIV.
[413] Matthew 16:26, NIV.
[414] Matthew 19:26, NIV.

far, I hope these reasons for the soul and the afterlife cause you to reflect more deeply on the matter.

I believe that I have refuted the atheist worldview in this book. But then again, we are hard-wired to know that God and the eternal realm exists (Ecclesiastes 3:11), so I really did not need to do that. An even better line of evidence for the reality of the soul is Christ's resurrection. If Christ was raised bodily from the dead – and he was – there is every reason to believe that his followers will be raised as well. We will discuss Christ's resurrection in more detail later in this chapter. Although I have already discussed the issue, let us examine in more detail the fact that we inherently realize that we are eternal beings designed to be in communion with God.

Bertrand Russell (1872-1970) was the most well-known atheist in the middle of the twentieth century. He once wrote, "The centre of me is always and eternally in terrible pain, a curious, wild pain, a searching for something, beyond what the world contains…"[415] That is a pretty serious problem for an atheist, but it was an honest admission on his part. The question that should immediately come to mind is this: Why was Russell constantly searching for something beyond this world, when the atheist recognizes nothing beyond this world? Interestingly, a contemporary – and philosophical opponent – of Russell's had the answer to his dilemma.

C.S. Lewis (1898-1963) is generally considered to have been the greatest Christian thinker of the twentieth century. Although he wrote many impressive books, *Mere Christianity* may have been his greatest work. Lewis explained why Russell, and countless others like him, searched for something beyond this world:

> Creatures are not born with desires unless satisfaction for those desires exists. A baby feels hunger: well, there is such a thing as food. A duckling wants to swim: well, there is such a thing as water. Men feel sexual desire: well, there is such a thing as sex. If I find in myself a desire which no experience in this world can satisfy, the most probable explanation is that I was made for another world.[416]

Like Bono and the rest of the boys from U2, Bertrand Russell never did find what he was looking for,[417] because he was always looking in the wrong place. (Actually, Bono and the other members of U2 are Christians, so they really have found the One who we all should be looking for.) Instead of looking to the spiritual world for the answer to his great need, Russell searched this world in

---

[415] Bertrand Russell wrote this in a letter to Lady Constance Malleson (1895-1975). She sold her letters from Bertrand Russell to McMaster University, where they are currently housed.

[416] C.S. Lewis, *Mere Christianity* (San Francisco, CA: HarperSanFrancisco, 2001), 136-137. 1952 Original Publication.

[417] U2, "I Still Haven't Found What I'm Looking For" (The Joshua Tree, 1987).

vain. But this world can never fulfill a need that is spiritual in nature. Russell had a God-shaped hole in his heart, which only God could fill. But since he refused to consider God, the hole remained open.

C.S. Lewis wrote that we were made for another world, but Paul beat him to the punch by two millennia when he revealed that, "Our citizenship is in heaven."[418] But then again, the psalmist beat Paul to the same point by a thousand years when he wrote, "I am a sojourner on the earth."[419] We exist physically here on the earth, but we are merely sojourners – or temporary travelers – in this world. Our true home is in Heaven.

As Grant Jeffrey pointed out so well, human history reveals the universal belief in the afterlife. We naturally recognize the transcendent in our lives. Michael Guillen, ABC News Science Editor and a physics professor at Harvard University, further comments on this fact:

> To our knowledge, no other creature on the planet senses the existence of an invisible realm – daily takes into account not just the here-and-now but the hereafter. No other creature believes in a Creator and worships him – builds magnificent cathedrals, composes brilliant symphonies, or founds universities and hospitals in his name.[420]

Gobekli Tepe, in modern-day Turkey, contains the world's oldest known temple complex. As far as anyone can tell, it predates anything built by the Egyptians or even the Sumerians. Yet it was amazingly complex in design. (We briefly looked at Gobekli Tepe in chapter six.) What does Gobekli Tepe teach us? Probably a lot of things, but in the context of this chapter it reveals to us that worship goes as far back as humanity itself. And why is that? Because we were hard-wired for worship, having eternity or Heaven 'set in our hearts' (Ecclesiastes 3:11). In fact, as a race we have worshiped the transcendent – although not always correctly – for so long that it has been suggested that instead of being scientifically classified as *Homo sapiens sapiens* ('wise wise man'), we instead should be known as *Homo sapiens religiosus* ('religious wise man').

The Resurrection of Christ

When it comes to the evidence for the soul and life after death, the resurrection of Jesus is the only evidence we really need. But like some professing Christians in the more liberal churches today, some followers of Christ in the early Church had a problem with Christ's resurrection: They could not get their minds around it, and therefore they rejected it. In a letter to the problem-plagued church in

---

[418] Philippians 3:20, NIV.
[419] Psalm 119:9, ESV.
[420] Michael Guillen, *Amazing Truths: How Science and the Bible Agree* (Grand Rapids, MI: Zondervan, 2015), 163.

Corinth, Paul wrote, "But if it is preached that Christ has been raised from the dead, how can some of you say that there is no resurrection of the dead?"[421] It seems incredible to me that any professing believer – whether in the first century, today, or anytime between – could doubt the resurrection of Christ. This belief is, after all, the linchpin of Christianity: Everything associated with the faith hinges upon it being true. Without Jesus being bodily raised from the dead, we might as well sleep in on a Sunday morning. (Although sleeping in every other day seems like a good plan to me.)

The problem with the Corinthian Christians – or, I should say, their major problem as there were many issues they struggled with – was their acceptance of a bad influence: Greco-Roman philosophy. Although they had one foot in the Church, they had the other foot in the philosophy and culture of the Corinthian-Gentile world. Aeschylus (523-456 BC), a Greek playwright, poet, and philosopher contemporaneous with Socrates, wrote, "When the dust has soaked up a man's blood – once he is dead – there is no resurrection." That idea, of course, included Christ's resurrection. Who did they trust more: The writings of Aeschylus, or the words of Jesus and his apostles? Although not every Corinthian believer doubted the resurrection of Jesus – Paul did, after all, write, "how can some of you say," not, "how can all of you say" – it showed a tendency to accept their former worldview over what should have been their new worldview. Once we profess Christ as Lord and Savior, everything changes. I remember how my thinking changed. I went from being a skeptical evolutionist to a believing creationist, and the world has never looked the same since.

Due to our fallen nature, combined with our upbringing which can be a powerful influence, it can be hard to stay focused on what we believe as Christians. I know people who came to Christ from a variety of other beliefs or religions, and sometimes they struggle against reverting back to their former way of thinking. One person, who grew up in a pseudo-Christian cult, told me that she sometimes finds herself thinking like a cult member on occasion, however briefly that might be. That cultic theology was embedded in her from a young age, and it can be hard to push that aside at times. People tend to think in the manner they were raised. Even later, when they realize they had been wrong, they can still fall back into that erroneous thinking merely out of habit. The Corinthian churches were predominately Gentile. Those believers were raised in the thinking of Thales, Aeschylus, Aristotle, and a host of others who knew nothing about the biblical God or the Judeo-Christian worldview. So, when they considered resurrection from the dead, some held fast to the Greek idea of no resurrection over what they should have now known to be true.

In Greek thinking, the body was considered inferior to the spiritual aspect of man. The Gnostics took this idea to an exaggerated level in the late first century and beyond, but the thought of the body being resurrected from the dead had long been considered abhorrent to the Greek mind. Therefore, when the earliest

---

[421] 1 Corinthians 15:12, NIV.

Christians spoke and wrote about bodily resurrection this conflicted thoroughly with Greek thought. The following Greco-Roman philosophers opposed the idea of bodily resurrection: Thales, Anaximander, Anaximenes, Parmenides, Pythagoras, Xenophanes, Heraclitus, Aeschylus, Anaxagoras, Zeno of Elea, Empedocles, Protagoras, Democritus, Aristotle, Epicurus, Zeno of Citium, and Lucretius. Except for Socrates and his student Plato, this is pretty much the 'Who's Who' list of Greco-Roman philosophy, and even Socrates and Plato were likely opposed to the idea of bodily resurrection as well. (At least as Christians understand bodily resurrection.)

Make no mistake about it, this kind of Greco-Roman thinking is still with us today, which is why we need apologetics training in the Church more than ever before. As believers we must be able to not only defend the biblical view on origins, which has been the bulk of our study, but we must also proclaim the evidence for Christ's resurrection as well.

Paul went on to correct the Corinthian believers who denied bodily resurrection: "But in fact Christ has been raised from the dead."[422] The resurrection of Jesus is the best evidence for life after death, as Christianity has the testimony of Jesus – who called himself back to life. We really do not need any other line of evidence beyond Christ's resurrection.

The 'Case for Christ' is the basis for Christian apologetics. Creation is important, but it is not the end goal. Eschatology (end times) is important, but it is not the end goal. Predestination versus free will is important, but it is not the end goal. How Christianity compares to the other religions of the world is important, but it is not the end goal. But Christ crucified for the sins of mankind, resurrected and victorious over death – that is the end goal of apologetics. There is no other topic that we can study more intently.

Although I love Lee Strobel's *The Case for Christ* – for a popular-level audience it seems to be the 'gold standard' among books that focus on Jesus' resurrection – I like Tim Chaffey's *In Defense of Easter* even more.[423] At this point in our study, I could re-hash the traditional 'Case for Christ,' but many authors have already done that very well. If you need to learn, or even review, the apologetics behind Jesus' resurrection, I recommend you begin with Chaffey's book and follow-up with Strobel's. Instead, let me present to you the apologetic that hinges on the idea that no human being could make up the Gospel message. It is an ancient argument for Christ's deity, yet one of the greatest apologists of the twentieth century utilized it as well. To understand this idea, we will examine the work of four writers: Isaiah, Paul, Tertullian, and C.S. Lewis.

Let me begin with Tertullian. Quintus Septimius Florens Tertullianus (AD 160-220) – now you know why we refer to him simply as 'Tertullian' – is often known for his 'pure faith' approach to knowing God. Tertullian said that what

---

[422] 1 Corinthians 15:20, NIV.
[423] The books on Christ's resurrection by Gary Habermas and Michael Licona are also some of the best on the subject.

made Christianity so trustworthy for him is that it is not based upon logic and reasoning, but rather it is based upon a supernatural worldview which in many ways runs counter to logic and reasoning, echoing the vast chasm between the infinite God and finite man. In short, Tertullian maintained the truth of the Gospel because mere men could not make up a story like that – it is so incredible that it had to come from God himself. The truth of the Gospel is not unreasonable, rather it is beyond reason.

Seven centuries before Christ, and nine centuries before Tertullian, Isaiah wrote about the vast chasm that exists between God and man: "For my thoughts are not your thoughts, neither are your ways my ways...For as the heavens are higher than the earth, so are my ways higher than your ways and my thoughts than your thoughts."[424] The difference in thought and action between God and man is not simply quantitative, but predominately qualitative. We think and act in certain ways, which are incomprehensibly different from how God thinks and acts. This explains, at least in part, why the words of Scripture can confound us at times.

Based upon this knowledge that God's ways are not our ways, and his thoughts are not our thoughts, Tertullian held fast to the belief that Christianity can only be explained by divine revelation. As a trained lawyer and an exceptional apologist, he knew the rules of logic and reasoning better than most people of his time. But accepting the Gospel message is an act of faith, not an act of logic. Tertullian also hung tightly to the words of Paul in 1 Corinthians. Read these words from Paul, keeping in mind Isaiah 55:8-9 and Tertullian's belief that the Gospel message could not be invented by mere men. My thoughts will follow in brackets:

> For Christ did not send me to baptize but to preach the gospel, and not with words of eloquent wisdom, lest the cross of Christ be emptied of its power. [Forget about appealing to the intellect, with 'eloquent wisdom.' As Tertullian and countless others have maintained, if a person is going to invent a story regarding salvation, the Gospel message would not be it.]
>
> For the word of the cross is folly to those who are perishing, but to us who are being saved it is the power of God. [Once again, from a merely human perspective the Gospel message does not seem reasonable; it is pure 'folly' to the unsaved. Nonetheless, this is God's clear plan for salvation.]
>
> For it is written, "I will destroy the wisdom of the wise, and the discernment of the discerning I will thwart." [The 'wise' – that is, those who reason as mere men, who do not understand the things of God – will never figure out Christ's sacrificial atonement for the sins of mankind. They cannot understand the

---

[424] Isaiah 55:8-9, ESV.

spiritual things of God, as their minds have not been regenerated through Christian conversion. In the next chapter of 1 Corinthians, Paul writes, "The person without the Spirit does not accept the things that come from the Spirit of God but considers them foolishness, and cannot understand them because they are discerned only through the Spirit."[425] This is why Augustine and Anselm of Canterbury were known for saying that in order to understand the things of God, we must first believe. They undoubtedly had these words from Solomon in mind as well: "The fear of the Lord is the beginning of knowledge."[426] If you want to understand God's plan for mankind, you must first believe in the God revealed in Scripture, by faith.]

Where is the one who is wise? Where is the scribe? Where is the debater of this age? Has not God made foolish the wisdom of the world? For since, in the wisdom of God, the world did not know God through wisdom, it pleased God through the folly of what we preach to save those who believe. For Jews demand signs [miracles and revelation] and Greeks seek wisdom [philosophy], but we preach Christ crucified, a stumbling block to Jews and folly to Gentiles, but to those who are called, both Jews and Greeks, Christ the power of God and the wisdom of God. [The 'wisdom of the world' cannot understand the Gospel message. Reason will not lead the unsaved person to conclude that the Gospel message is true. Instead, fallen man will seek different stories to explain God's existence and nature, and how we can relate to whatever version of God people choose for themselves. This explains why mankind has hundreds of different religions and philosophies. In the end, however, there are only two major worldviews: The God-centered (biblical) worldview, and the man-centered (secular) worldview. None of the humanly devised religions or philosophies would ever think up something as 'strange' as the message of the cross – it is truly a stumbling block to the Jews and foolishness to the Greeks.]

For the foolishness of God is wiser than men, and the weakness of God is stronger than men.[427] [Consider these words from Solomon: "Trust in the Lord with all your heart and lean not on your own understanding; in all your ways submit to him, and he will make your paths straight."[428] This says it all.]

---

[425] 1 Corinthians 2:13, NIV.
[426] Proverbs 1:7, NIV.
[427] 1 Corinthians 1:17-25, ESV.
[428] Proverbs 3:5-6, NIV.

Although Tertullian was not opposed to using logic and reasoning when the situation demanded it – he was, after all, a lawyer by training who most certainly possessed critical thinking skills – he nonetheless rejected the idea that one could mix Christian faith with Greco-Roman philosophy. (This was unlike his apologetic predecessor Justin Martyr, who was the first of the great apologists to live after the time of the apostles. Justin was well-known for his synthesis of Christian theology and Greek logic, however strained it may have been at times.) Tertullian feared that pagan philosophy, if used incorrectly, could distort the message of Christ. Unfortunately, history has shown us that Tertullian was correct. Along this line, he is perhaps best known for his question, "What has Jerusalem to do with Athens, the Church with the Academy, the Christian with the heretic?" Tertullian was impressing upon his readers that there should be no attempt to integrate the doctrines of the Christian faith with the ideas of fallen man, regardless of how logical and well-reasoned they may seem to be. Jerusalem, representing revealed religion, can have no relationship with Athens, which represents purely human philosophy or speculation. Likewise, the Church cannot cooperate with Plato's Academy (secular-based academia today), as the Christian and the pagan are light years apart in their thinking. For Tertullian, Christianity and pagan philosophy simply do not mix, and any attempt at synthesizing the two would result in nothing less than the watering down of Christian doctrine.

In the twentieth century, no less than C.S. Lewis – almost universally considered to be the greatest apologist of that century – shared the same belief as Tertullian. (And Isaiah, Solomon, and Paul, for that matter.) Lewis wrote:

> Reality, in fact, is usually something you could not have guessed. That is one of the reasons I believe Christianity. It is a religion you could not have guessed. If it offered us just the kind of universe we had always expected, I should feel we were making it up. But, in fact, it is not the sort of thing anyone would have made up. It has just that queer twist about it that real things have.[429]

This is exactly the point that Tertullian made seventeen centuries earlier. The Gospel message must be true, because no one would ever make up something even close to it. This is an apologetic that works for me, but admittedly it may not work for everyone. If you are interested in further exploring this approach to the 'Case for Christ,' I highly recommend Tom Gilson's *Too Good to be False*. This approach to Christ's deity also dovetails perfectly with my preferred apologetic method.

Several apologetic methods have been devised over the centuries. The classical method begins by making the case for God's existence in general, and

---

[429] Lewis, 41-42.

then concludes with the historical evidence for Christianity. (This explains why it is often referred to as the 'two-step approach.') My dissertation focused on the first step in the classical method, in which I utilized science and philosophy as pointers to God. This method also formed the basis for my first book, *Worldviews in Collision*. Many of the greatest defenders of the faith were classical apologists, such as Justin Martyr, Augustine, Anselm, Thomas Aquinas, William Paley, B.B. Warfield, Norman Geisler, and R. C. Sproul. It appears that I am in good company!

The evidential method mixes everything (science, philosophy, history, biblical theology, world religions, etcetera) into a 'one-stop' case for the Christian worldview. (Not surprisingly, it is referred to as the 'one-step approach.') Brian Huffling, the director of the Ph.D. program and an associate professor of philosophy and theology at Southern Evangelical Seminary, describes the rationale behind the evidential approach:

> Evidential apologists avoid an attempt to demonstrate that God exists. Some do this because they don't think natural theology is possible; others think it is simply easier to start with the biblical case. They jump straight to evidence's for showing that Christianity is true from fields such as history and archaeology. To them, this bypasses difficult philosophical arguments and objections. People are ordinarily more prone to understanding history and the like. The thinking here is: if we can show the Bible to be reliable and that Jesus was raised from the dead, then a reasonable person will be convinced that Christianity is true. Such would include the existence of God.[430]

Presuppositional apologetics maintains that everyone already knows that God exists – a point which I totally agree with – and the only reason that we can even think through these theological-philosophical issues in the first place is because God has designed us with the gifts of logic and reason, which can only be truly implemented in the context of scriptural truth. Presuppositional apologists' reason from the Bible, whereas evidence-based apologists' (classical and evidential) reason to the Bible. It is never that black-and-white in real life, but in general this is a fair assessment between the two approaches. There are several other methods as well, including the other one I favor: Rational fideism. However, rational fideism should not be confused with traditional, or pure, fideism.

Traditional fideism maintains that we believe simply by faith. (Fideism comes from the Latin word *fides*, which means faith.) In fideism, faith is elevated far above reason. In fact, evidence is oftentimes relegated to the background, if not

---

[430] Brian Huffling, "Apologetic Methods and a Case for Classical Apologetics." https://ses.edu/apologetic-methods-and-a-case-for-classical-apologetics/

pushed to the sidelines altogether. But rational fideism is a different animal.

In rational fideism, faith takes over where reason leaves off. For example, I believe in the triune nature of God. This is the Christian belief that God is one, but existing in three persons. No human mind can wrap itself around this doctrine, but I believe it by faith. Looking at nature does not reveal God's triune nature to me, but nature does plainly reveal God's existence (Romans 1:20). Belief in God is rational, but belief in the Trinity is not something that I would arrive at through normal observation or logic. Instead, divine revelation is how I know this doctrine is true. Since Scripture gives me countless reasons to believe in God's triune nature, I accept by faith this doctrine. I do not believe Scripture is true because I simply want Scripture to be true, or because I presuppose it to be true. Rather, Scripture has proven itself to me through fulfilled prophecy, scientific accuracy hundreds if not thousands of years prior to modern verification, and the fact that it speaks to my soul in a way that no other collection of books could even come close to doing. Therefore, God exists (reason), and Scripture reveals that God is triune in nature (faith). Reason and faith work together. When it comes to accepting God's triune nature, reason can take me only so far – and then my well-reasoned faith in Scripture takes me the rest of the way.

Therefore, I believe that making the case for God's existence first is the best foundation. Along with presuppositional apologists, I agree that all people inherently know the Creator exists. The problem is this: People are way too good at deeply burying that inherent knowledge, whether inadvertently or even intentionally. Therefore, we need to first establish God's existence in general before moving on to the evidence for Christian theism specifically. After all, with God all things are possible (Matthew 19:26), be it miracles, fulfilled prophecies, or the resurrection of Christ. God's existence is the foundation for all things supernatural, be it Heaven, Hell, angels, demons, or the immortal soul.

Along the way, we need to always remember that reason can take us only so far: Reason and faith work together. Where reason leaves off, faith takes over. However, faith cannot take over if the foundation is unreasonable. And the foundation of apologetics, in my opinion, is God's existence. Classical apologetics is the oldest of the apologetic methods for a reason: It works. And rational fideism stresses the positive relationship between faith and reason. This is a powerful combination of apologetic methods.

Besides the traditional apologetic for Christ's deity – which is laid out so well by Chaffey, Strobel, and numerous others – and the belief that the Gospel message could not be invented by fallen man, I hold to the power of fulfilled prophecy as well. For example, the Old Testament prophet Isaiah lived seven centuries before Christ, yet he foretold of the Lord Jesus Christ with amazing accuracy and insight. Through the power of the Holy Spirit, Isaiah was able to peer down the corridor of time and see Christ himself. Perhaps no other section of Isaiah's writing demonstrates this fact better than the 'Suffering Servant' passage in chapter fifty-three.

Isaiah wrote that Messiah will be a sin offering, a sacrificial 'Passover lamb.' This is one of the most amazing passages in the Old Testament concerning the coming Messiah. Isaiah 53 is of special interest to the Christian believer, as these verses were written over seven centuries before the birth of Christ, yet Jesus – and Jesus alone – fulfills this passage perfectly.

This amazing chapter is found in Jewish Bibles today, though it is generally left out of the weekly synagogue readings. Many modern rabbis say that the suffering described by Isaiah are those of the nation of Israel. However, most ancient rabbis said that the passage referred to Messiah's sufferings, not to those of Israel as a nation. This includes ancient rabbinical commentators from the Babylonian Talmud, Midrash Ruth Rabbah, Zohar, and even the great Jewish thinker Rabbi Moses Maimonides.

Isaiah 53 cannot refer to the nation of Israel, or any mere human being, but only to Jesus – who was, and is, fully God and fully man at the same time. This conclusion is based upon several points. First, the servant of Isaiah 53 is an innocent and guiltless sufferer, yet Israel is never described as sinless. Second, Isaiah said that it pleased the LORD to bruise the servant. Has the awful treatment of the Jewish people really been God's pleasure? Clearly the answer is no. Third, the person mentioned in this passage suffers silently and willingly. Yet all people, Israelites included, complain when they suffer. Only Christ, in the New Testament, goes to his death quietly and in full submission to God. Fourth, the figure described in Isaiah 53 suffers, dies, and rises again to atone for his people's sins. Isaiah 53 describes a sinless and perfect 'sacrificial lamb' who takes upon himself the sins of others so that they might be forgiven. The terrible suffering of the Jewish people does not in any way atone for the sins of the world. Isaiah 53 speaks of one who suffers and dies to provide a legal payment for sin, so that others can be forgiven. This cannot be true of the Jewish people collectively, or of any individual man. Fifth, the prophet speaking is Isaiah himself, who says the sufferer was punished for "the transgression of my people" (verse eight). The people of Isaiah are Israel, therefore the servant suffered for Israel, and therefore was not Israel herself. Sixth, the figure of Isaiah 53 dies and is buried, according to verses eight and nine. The people of Israel have never died in their entirety. They may have come close during the Holocaust, but even a significant number of the Jewish people survived that horrible episode of their history. Seventh, if Isaiah 53 cannot refer to Israel, can the passage refer to Isaiah himself? Isaiah said he was a sinful man of unclean lips, therefore Isaiah could not die to atone for our sins. Nor could it have been Jeremiah or Moses. Isaiah, Jeremiah, and Moses were all prophets who gave us a glimpse of what Messiah would be like, but none fit the description of Isaiah 53, for all were sinners and fallen in nature. Beyond any doubt, Isaiah is referring to Jesus, the 'Lamb of God who takes away the sins of the world' (John 1:29).

Table 9.1

Evidence for the Soul and the Afterlife

THE ATHEIST WORLDVIEW HAS BEEN THOROUGHLY REFUTED.

PEOPLE INHERENTLY UNDERSTAND THAT WE WERE CREATED FOR ETERNITY.

THE RESURRECTION OF JESUS CHRIST

SHOWS US THAT LIFE CONTINUES BEYOND THE GRAVE.

To summarize, the atheist position on the soul and the afterlife is thoroughly bankrupt, which leaves us with the transcendent view of life after death. The single best evidence for Heaven is Christ's resurrection, which apologists have been defending since that event happened. We are destined for Heaven, provided we accept Christ's gift of salvation. Do not miss the ultimate message of Romans: "If you declare with your mouth, "Jesus is Lord," and believe in your heart that God raised him from the dead, you will be saved. For it is with your heart that you believe and are justified, and it is with your mouth that you profess your faith and are saved."[431] Is it really that simple? Yes, it is. Do not be deceived by those who profess a different – and unnecessarily complicated – Gospel.

Heaven: The Glorious Kingdom

Let us now turn our attention to the glorious kingdom to come, Heaven. I want to make a few points regarding Heaven. First, in Heaven we will be healed and repaired. The Holy Spirit, through Paul, reveals to us that in the world to come we will have indestructible bodies that will never know sickness or grow weak: "For our dying bodies must be transformed into bodies that will never die; our mortal bodies must be transformed into immortal bodies."[432] This is a verse that offers great hope.

After my heart attack, it was determined that I suffered from an arrythmia (electrical condition) that required a pacemaker-defibrillator for my protection. Although my heart paces itself without the use of the pacemaker, the defibrillator is there in case my heart goes into a deadly rhythm and needs to be shocked back into the proper beating pattern. Numerous people have said to me, "It is just like having an emergency team in your chest, ready to save your life at any moment." I know they mean well, but trust me, having a pacemaker-defibrillator feels more like a curse than a blessing much of the time. (The shock delivered to the heart can take a strong man to his knees. It has only happened to me once, so far, and it is most unpleasant.) I have had the device implanted in my upper chest for two years now, and not a day goes by that I do not say to myself, "I miss the old days, when I was not a cardiac patient and did not have this thing stuck in my chest."

---

[431] Romans 10:9-10, NIV.
[432] 1 Corinthians 15:53, NIV.

When I eventually pass, the first thing I will do in Heaven is feel the left side of my chest – and the device will not be there. To be healed and repaired will be glorious beyond imagination.

Second, never count anyone out of Heaven. Until we are there, we simply do not know who inhabits this glorious kingdom. I briefly discussed Friedrich Nietzsche in chapter four. He was a hardcore atheist who gave no quarter to Christians. If you asked the average believer sitting in the pews today, "Where is Friedrich Nietzsche now?" the answer would probably be, "In Hell." Although it does not look good for Nietzsche, the truth of the matter is we simply do not know where he is eternally. (Of course, my atheist friends would deny that he is anywhere today, for they say that when he died in 1900 his life ended at that very moment. I could not disagree more.) It is true that he could be in Hell, eternally separated from God. But for all we know, he could be in Heaven with the Lord himself. I really hope that is the case.

Arthur Francis Green is one of my favorite apologists. He is a true scholar who is exceptionally well read in philosophy, and he understands better than most how secular thinking has impacted the world over the centuries. His book *When Fables Fall: Unmasking the Lies of Distorted Science, Secularism, and Humanism* is masterful in this area. Regarding Nietzsche, Green once said, "I hope to meet Friedrich Nietzsche in Heaven someday, for he will have been healed and repaired."[433] The first time I heard this, I was stirred in my soul. Forget about theology degrees, winning debates with atheists, and making an apologetic point. This is what Christianity is supposed to be about: Holding out hope for other people, pointing them to a life in Christ and the world to come. We should all have that Christian attitude of hope for the lost and broken.

One of the reasons that I included this chapter after addressing the eternal fate of the unevangelized (chapter eight) is that I contend we will be amazed over who is in Heaven. I am convinced that we will meet Hindu's, Buddhist's, Muslims, and a host of other people who did not recognize Christ as their Lord and Savior for most of their earthly life – maybe even up to the point of irreversible death. That makes me a heretic in a lot of Christian circles, but please hear me out first. In chapter eight I discussed universal presentation, inclusivism, and the 'final option' view of post-mortem salvation. I believe it is possible that these views might work together in such a way that many of the unevangelized could become followers of Christ in what we would deem the 'last possible moment' before final (irreversible) death.

I believe that everyone who enters Heaven does so only through Christ. Jesus made it clear that, "I am the way and the truth and the life. No one comes to the Father except through me,"[434] and Luke plainly wrote that, "Salvation is found

---

[433] Arthur Francis Green, "When Fables Fall." Unmasking Fables, Promoting Truth DVD (Creation Ministries International). https://usstore.creation.com/product/401-unmasking-fables-promoting-truth-dvd
[434] John 14:6, NIV.

in no one else, for there is no other name under heaven given to mankind by which we must be saved."[435] If a Hindu, a Buddhist, or a Muslim is in Heaven, it is because they accepted Jesus Christ as their Lord and Savior at some point prior to final death. Maybe it was in the final moments of consciousness, or maybe it was at the interface between life and death – unconscious and unresponsive to their fellow man, but not yet dead. Do not rule that possibility out. (I recommend reading, or re-reading, about universal presentation, inclusivism, and post-mortem salvation in chapter eight.)

Although this idea sounds like it is leading to universalism, it does not have to. Admittedly, it could for some people, but not for me. Hell is real, and people can choose to reside there. (Or be annihilated. That debate is still roaring.) Jesus talked about Hell as though it were a real destination, and not just some imaginary concept intended to scare people into good behavior. People have free will, and they will choose to be with, or apart from, God for eternity. I cannot imagine someone looking the Creator of the universe in the eye and saying, "Forget you." But in a universe in which both free will and the fall of man exists, that could conceivably happen.

I strongly suspect that more people will make it into Heaven then many Christians realize, or will admit in some cases. I say 'admit in some cases' because in a few internet debates I have been involved in, some professing Christians seem to be convinced that Hell will be filled to the brim with all sorts of people. And the verse they always rely upon? "But small is the gate and narrow the road that leads to life, and only a few find it."[436] But what do these same folks do with Revelation 7? I love these words from John:

> After this I looked, and behold, a great multitude that no one could number, from every nation, from all tribes and peoples and languages, standing before the throne and before the Lamb, clothed in white robes, with palm branches in their hands, and crying out with a loud voice, "Salvation belongs to our God who sits on the throne, and to the Lamb!"[437]

A multitude of souls who had once come from every nation, tribe, people, and language group is nothing short of amazing. But not every nation, tribe, people, or language group has historically been associated with Christianity or the Bible. Or even one Creator-God, for that matter. Are these non-biblical people included in the Revelation 7 vision? I see no reason why they are not. For example, Hindu's will be in Heaven, but not because they were Hindu's while alive on the earth. At some point between life and irreversible death, each of these Hindu's who reside in Heaven will have recognized Jesus for who he really is. I am not

---

[435] Acts 4:12, NIV.
[436] Matthew 7:14, NIV.
[437] Revelation 7:9-10, ESV.

saying that every Hindu who has ever lived will automatically be in Heaven someday. But I have no doubt that many will. This belief of mine will sound heretical to those Christians who are extremely conservative, and likewise it will sound arrogant to those from non-Christian religions who tend toward universalism. But I believe I am being biblically sound on this matter: Salvation is found in Christ alone (John 14:6; Acts 4:12), God wishes for all to be saved (1 Timothy 2:1-4), God is just (2 Thessalonians 1:6; Revelation 15:3), and God loves the entire world (John 3:16). Put all these points together, and then reflect upon it with an open heart before you consider labeling me a 'false teacher.' (I realize many of you will happily agree with me on this matter. I have met many believers who do.)

Third, death is all in God's timing. "Your eyes saw my unformed substance; in your book were written, every one of them, the days that were formed for me, when as yet there was none of them."[438] I would brush aside a verse like this in the past, for my attitude was always, "With enough exercise, proper diet, and self-care, anyone can extend their lives a certain amount, save an accident or developing an aggressive cancer." That was the 'exercise junkie' philosophy, but the problem is it does not always work. God is in control, not me, and the number of days allotted to me is determined by the Creator. Exercise is great. I feel better because of it, and it has helped me both physically and mentally over the course of my life. But my lifespan is determined by God: "A person's days are determined; you have decreed the number of his months and have set limits he cannot exceed."[439] Yes, continue to exercise – or start exercising if you have avoided it so far – but do not be overly confident and think you will live a super-healthy life into old age because of it. Jim Fix (1932-1984) was an American runner who wrote the 1977 best-selling book *The Complete Book of Running*. Fixx played a major role in jumpstarting America's fitness craze by popularizing running. He was thin, and he seemed to be in great shape. But in his early fifties, he died of a massive heart attack. No one saw that coming – except God, who already knew the exact time that he would die.

Prior to my heart attack, I made plans for my life. There is nothing wrong with that. In fact, making plans is a good thing. However, my plans never included the phrase 'God willing.' That was the mistake I made. Reading through the book of James made me realize that. I love this book, and I have taught from it in Sunday morning classes and small group studies several times. But these verses never fully impacted me until after my heart attack:

> Now listen, you who say, "Today or tomorrow we will go to this or that city, spend a year there, carry on business and make money." Why, you do not even know what will happen tomorrow. What is your life? You are a mist that appears for a little while

---

[438] Psalm 139:16, ESV.
[439] Job 14:5, NIV.

and then vanishes. Instead, you ought to say, "If it is the Lord's will, we will live and do this or that.[440]

'If it is the Lord's will' is a phrase we all need to keep in the forefront of our minds. Make plans for your life, but keep in mind that your plans are subject to God's will for your life. You have a limited amount of control over your life.

Fourth, and finally, salvation comes through declaring Christ as Lord and Savior – and by doing nothing else. Paul wrote, "If you declare with your mouth, "Jesus is Lord," and believe in your heart that God raised him from the dead, you will be saved."[441] In an online course that I took several years ago, a young man confidently declared in the class chat room that to be saved, one must believe in a young earth. I will let some things slide on by, but that kind of bad theology must be addressed. It took some convincing before this young man eventually saw the error in his thinking. One of the instructors, a world-class creationist speaker and author, also told the young man that an old earth believer is every bit as saved as a young earth believer. It seemed to be more of an emotional roadblock than an intellectual obstacle for this zealous young student. Some recent creationists still hold to this idea, unfortunately. The Romans 10:9 profession is what makes a person a saved believer. You can be a recent creationist, an old earth creationist, a theistic evolutionist, a gap theorist, or hold to any other position on origins and still be saved if you profess Christ as Lord and Savior. Likewise, you can hold to any of the various eschatological (end times) scenarios and be just as saved as anyone else. You can be a Calvinist or an Arminian and be truly saved. (As much as young earth and old earth creationists love to fight, their battles can seem mild when compared to the all-out war that Calvinists and Arminians engage in!) My main point: Do not add to Romans 10:9, as this verse, like all others, was never intended for a later addition (Revelation 22:18-19).

There are two books that I highly recommend concerning death and the afterlife. Clay Jones' *Immortal: How the Fear of Death Drives Us and What We Can Do About It* is a fantastic resource for those who struggle with the idea of leaving this world behind. In similar fashion, Michael Wittmer's *The Last Enemy* is an encouraging read for those who know that death will bring great things, but still wrestle with the idea of leaving behind family, friends, and important life projects. Even the most devout among Christ's followers may find themselves clinging to this life, and this is not unexpected. All people, regardless of their worldview, desire to remain in this life to varying degrees.

---

[440] James 4:13-15, NIV.
[441] Romans 10:9, NIV.

Concluding Thoughts

We began this chapter with a timeless question from Job: "When a man dies, will he live again?"[442] Jesus answered that question much later in history. Jesus said to Martha, the sister of his good friend Lazarus, "I am the resurrection and the life. The one who believes in me will live, even though they die."[443] Our earthly lives are a gift from God, but as C.S. Lewis noted in the middle of the previous century, we were "made for another world."[444] Our earthly lives can be wonderful at times, but Heaven was always meant to be the endgame. Trust in Christ, and you will be certain that Heaven is your final destination.

Let us now summarize the history of the origins debate, and recap some of the major points from chapters five through nine as well. It is my sincere hope that the reader has discovered much to chew on, whether you agree with me or not on the various points discussed.

---

[442] Job 14:14, NIV.
[443] John 11:25, NIV.
[444] Lewis, 137.

CONCLUSION

Does Any of This Really Matter?

The question we must ask ourselves is this: "Does any of this really matter?" We can learn a lot about the history of the origins debate, about the evidence for the Creator and the reasons why evolutionism is a bad philosophy, about differing approaches to solving the problem of suffering and evil, and about death and the afterlife – and who may or may not go to Heaven when this life is over – but what difference does this knowledge make in our lives? Can the information contained within the pages of this book really change a life? Yes, I know it can – it changed my life, for starters. Before we examine the relevance of this book, though, let us first review what we have learned in this study.

The Long History of the Origins Debate

After examining the history of the origins debate, we may come to some solid conclusions on the matter. First, this is a spiritual battle. Make no mistake about it, this is spiritual warfare in every sense of the term. Although we cannot blame evolutionary teachings solely on Satan and his followers, it seems very reasonable that he is behind this destructive philosophy in some way, shape, or form. There may be no better way to lead people astray than to convince them that there is no God, that we are nothing more than a cosmic accident. He may have even tried to convince himself that he evolved from the primordial waters described in both Genesis 1:2 and some of the ancient pagan cosmologies of Egypt and Mesopotamia. Satan might find it more comforting to believe that he evolved, as opposed to being created by the one whom he loathes.

Evolutionary teachings go way back into history, at least to the time of the ancient Greeks and are even found in proto-form in some earlier civilizations (Babylonians, Egyptians, Hindu Brahmins). Although admittedly speculative in nature, it is reasonable to consider that evolutionary teachings went back even further into history, to the early post-Flood era and even into the pre-Flood world. Since Satan has been able to successfully use evolutionary ideas from the time of the Greeks and Babylonians to the present, then why not believe that he used them at an even earlier time? Evolutionary teachings are truly a well-worn 'doctrine of demons.'

The ancient Hebrews were the exception to evolutionary thinking, however. Their worldview could be summed up in the first verse written by Moses: "In the beginning God created the heavens and the earth."[445] All other ancient cultures subscribed to some form of evolutionism, making the Hebrews' the only real creationists in antiquity.

In the ancient world that we know about–that is, the world after the dispersion at Babel–it was not until the time of Democritus' atomist school in Greece that

---

[445] Genesis 1:1, NIV.

full-blown atheism became a worldview. Prior to that time, it seems that evolutionary teachings were combined with some sort of belief in God or the gods, making it an early form of theistic or even polytheistic evolutionism. Rome continued the evolutionary thinking of its cultural predecessor, Greece. Evolutionism in the ancient world, through its devaluing of human life, was responsible in part for moral failures such as racism, slavery, abortion, infanticide, and a host of other social ills.

By the time of the early Church, the debate over origins was already ancient. Much of the Bible addresses the false teachings found not only within the professing body of believers but also from the non-believing world as well. The issue of origins was important to the earliest followers of Christ because they could see the impact that wrong ideas about God and creation could have on people. One example of this is Paul's Gospel address to the Epicurean and Stoic philosophers in Athens. This account from Acts 17 is the template for sharing Christ across worldviews. It is apologetics at its finest.

Augustine is widely considered to be the greatest Christian thinker after the time of the apostles. His view on creation was unorthodox in the eyes of some, however, as it is believed he favored an instantaneous creation of all matter-energy either prior to, or at the beginning of, Day One. This was followed by six days devoted to fashioning that matter into the world and universe we see today.[446] He was, nonetheless, a special creationist who denied the Greco-Roman evolutionism of his day.

The debate over the age of the earth is not a new issue, but rather extends back to the time of the early Church and even beyond. Like Darwinism today, Greco-Roman evolutionism required vast eons of time to be considered plausible. Without 'deep time,' evolutionism is nothing more than a failed hypothesis, and hence the push for an increasingly ancient universe and earth.

In the Middle Ages the Church continued to uphold the veracity of the Genesis account of creation, even though pagan philosophy was beginning to find its way into Christian scholarship during this time. With the arrival of the Renaissance around 1400 came an even greater emphasis upon ancient pagan ideas of origins. Humanism began to take hold at this time, and it became even more common during the Enlightenment or 'Age of Reason' beginning in 1600.

Although unwittingly, Thomas Aquinas introduced humanistic ideas into the Church through his reliance upon the ideas of Aristotle. A few centuries later, Sir Francis Bacon sought to remove both Greco-Roman philosophy and Scripture from empirical (observational-testable) science. Skeptics of Genesis creation seized the moment, however, and attempted to further remove both philosophy and Scripture from historical (origins) science as well. This humanistic trend has continued to this very day.

The Enlightenment was a period in which the Bible was attacked especially hard. Humanistic ideas fostered by blatantly rebellious men challenged the

---

[446] That is, the world we see today accounting for the effects of the Noahic Flood.

scriptures at every turn. This 'Age of Reason' set the stage for many opponents of Christianity, eventually paving the way for Charles Darwin's evolutionary ideas in the nineteenth century. Interestingly, the greatest humanistic influence in the Enlightenment may have been none other than Charles Darwin's grandfather, Erasmus Darwin. Based upon his prodigious work, it seems clear that he was the 'real Darwin' behind Darwinism. Nonetheless, it was Charles who successfully packaged the theory of biological evolution and sold it to a world that was waiting for some type of scientific verification in its quest to dismiss the Creator. Even today, most people know who Charles Darwin was, but they have no idea about the work of his grandfather or the other naturalists even in Charles' time.

Many readers may believe that I spent far too little time discussing Charles Darwin in this book. That was intentional, however. I want the reader to understand that Charles Darwin is only one link in the chain of evolutionary thinking. Far too much emphasis has been placed upon the younger Darwin when it comes to the philosophy of evolutionism, but we need to see that his role was much more limited than most people assume it was. He just happened to be in the right place, at the right time. He stood upon the shoulders of many before him, as we have seen. For an excellent study of Charles Darwin's religious ideas, consider obtaining David Herbert's *Charles Darwin's Religious Views*. It is a fascinating look at the ideas that shaped Darwin's worldview.

Evolutionary ideas took a wrong turn shortly after the time of Darwinism's rise to prominence. Along with Darwin, the ideas of Ernst Haeckel, Friedrich Nietzsche, and Francis Galton set in motion new ideas about supposed racial superiority. Although feelings of racial superiority had always been present in the world, there was now a new ally in the quest for this belief and the evils that accompany it: Science. Or, more accurately, science as understood through the lens of a very ancient philosophy revised for a new era (Darwinism). Eugenics was the new science of how some individuals and even whole people groups are inferior to others based upon physical characteristics. Unfortunately, in the early twentieth century eugenic ideas inspired a truly evil dictator from central Europe in his quest to rid the world of those who he deemed inferior and unworthy of contributing to the future human genome. Even today, when people are asked the question, "Who is the evilest man who ever lived?" it is likely that the name Adolf Hitler will be given as the answer.

Theistic evolutionism gained prominence in the Church shortly after the time of Darwin. Many theologians were quick to integrate evolutionary ideas into Genesis 1-2, whereas the scientists of the time were often opposed to Darwin's ideas. That is not what we would have expected.

Although young earth creationism is considered by some to have originated with George McCready Price early in the twentieth century, it was really the 1961 publication of *The Genesis Flood* by Henry Morris and John Whitcomb that launched the movement. However, the movement was nothing new even then, as many Church Fathers held to a literal six-day creation in the recent past as well as a global Flood in the days of Noah. Morris and Whitcomb simply reignited the fire for the recent creation account of origins that had been attacked hard since the time of the Enlightenment.

Throughout the twentieth century old earth creationism gained strength among believers, while a mystical-pantheistic form of evolutionism became common outside the walls of the Church. (And, sadly, even this view is embraced by some who profess Christ.) Both old earth creationism and pantheistic evolutionism pushed the recent creation view to the sidelines, a trend that began in the Enlightenment and continues today.

The science fiction and popular science writers of the twentieth century promoted evolutionary ideas through their writings, which were widely received. From H.G. Wells to Carl Sagan, evolutionism had its 'day in the sun' with these beloved writers. The extraterrestrial hypothesis gained a lot of momentum from them as well, although the idea that there are multiple inhabited worlds created by God goes back centuries.

The 1990's saw the rise of the Intelligent Design Movement. On one hand, ID has been a boon to creationism, as it has captured the interest of people who might not otherwise be interested in examining resources that appeal to divine design. On the other hand, it ignores the identity of the Intelligent Designer himself, Jesus Christ. ID is useful, but it merely introduces unbelievers to the origins debate. The Gospel message must conclude any evangelistic encounter.

Fortunately, creationist organizations have done much to promote biblical creationism worldwide. Answers in Genesis Ministries captured the attention of untold thousands throughout the world – both believers and unbelievers – with their Creation Museum and Ark Encounter. Likewise, the Institute for Creation Research recently unveiled their Discovery Center for Science & Earth History, while Creation Ministries International and the Creation Research Society continue to promote creationism both nationally and internationally. Besides the efforts of these major ministries, countless regional and local ministries are answering the call to spread the message of biblical creationism. Midwest Creation Fellowship, which serves the greater Chicagoland area, is one of the more successful regional ministries while Helmut Welke's Quad Cities Creation Science Association, on the central Illinois-Iowa border, is one shining example of a local ministry.

Not to be outdone by their young earth counterparts, Reasons to Believe promotes old earth creationism while BioLogos encourages believers to consider theistic evolutionism. If anything, these differing creationist ministries reveal that believers have divided into several different 'camps' concerning origins. It also forces the believer to sharpen his or her apologetic concerning origins.

When you examine the long history of evolutionism, which extends from the ancient world to the present, it becomes clear that this is an ideology that flies in the face of both biblical and scientific truth. This false view of origins is truly a 'doctrine of demons' (1 Timothy 4:1) which can hinder one from accepting the Gospel message.

Building the Case for Christianity

The remainder of the book addressed the evidence for the Creator, as well as reasons why recent creationism makes sense. The final chapters, which were more theological in nature, defended Christianity against common skeptical objections to the faith. It is good to know the history behind the origins debate, but we must go further and know why we can trust the biblical text.

In chapter five we examined the evidence against evolutionism as we built the 'Case for a Creator.' The unbeliever, like all people, inherently knows that God exists (Ecclesiastes 3:11; Romans 1:20; 2:14-15), yet they suppress this knowledge (Romans 1:18-32; 1 Corinthians 2:14). Therefore, evidence may play a limited role in sharing your faith with a skeptic, for regardless of how convincing the science, history, and logic may be in the case for biblical creationism – and it is overwhelming – the unbeliever may simply ignore it. However, there are times when evidence is of great value: It certainly worked for me many years ago, when I was a seeker of truth. We examined the biblical evidence of a universe with a beginning, a universe that demonstrates intelligent design in every way, and the cohesiveness of science and Scripture.

In chapter six, we examined the recent creation view. We explored the age of the creation, including solutions for a young universe that is billions of light years in diameter, as well as radiometric dating and fossils. The notion that the days of creation are literal 24-hour days is also something that should not be readily dismissed by either skeptics or old earth believers. We also addressed the recent creation of mankind, who was made in God's image.

In chapter seven we examined the problem of suffering and evil. Among other things, we considered Canaanite destruction, slavery in the Old and New Testaments, and a host of other skeptical objections to the faith. We concluded the chapter with a look at the 'bad fruits' of evolutionism, contrasting them with the 'good fruits' of the creationist worldview.

In chapter eight we examined the issue of eternal justice, in the form of the skeptical claim that Christians automatically condemn to Hell those who never even heard the message of the Gospel while still alive on the earth. This is not true, however, as Christians hold to several ideas that allow for the possibility of the unevangelized to be reconciled to God when this life is over. Agnosticism, exclusivism, inclusivism, post-mortem salvation, accessibilism, and universal presentation are some of the more common evangelical positions, while universalism, pluralism, purgatory, and reincarnation are maintained by other professing Christians as well, albeit these views contain serious problems. Lastly, we examined the issue of annihilationism versus eternal conscious torment, which is the traditional view of Hell. To say that all Christians are overly quick to condemn the unevangelized to Hell is simply wrong.

Finally, in chapter nine we discussed the evidence for the soul and the afterlife. There are two major views concerning this matter: The atheist view, which maintains that life ends at the grave, and the transcendent view, which proclaims the reality of the immortal soul and the continuation of life after death. Evolutionism is a worldview focused on death, as the central idea behind this philosophy is that currently living life forms must eventually make way for future – and 'evolutionarily more advanced' – life forms. This idea is in sharp contrast to creationism, which is the worldview of life.

Does Any of This Really Matter?

As we conclude our study, some may rightly ask the question, "Does any of this even matter?" What difference do these topics really make in a person's life? It turns out that it can make a huge difference. Over thirty-two years ago, as a biblical skeptic I wrestled with what I perceived to be a lack of 'cosmic justice.' I was a radiation therapy student in 1988, and I treated many children for cancer in that year. I witnessed a ten-year old boy who I had grown very fond of succumb to a brain tumor. As a result, I was angry with God (massive understatement) and at that point I could have gone the route of atheism. I seriously thought about it, as I wanted nothing to do with a God who allowed children to suffer and die. During that time, I was as close to embracing atheism as I could possibly get, without taking the final step. It was as if I could see myself standing partway through the opening to a world without God, ready to take the final step through that imaginary entrance but hesitant to do so. Why was I reluctant?

One thing kept me from taking that final step: My studies in biology and physics convinced me that God is real. I was challenged by my biological studies professor to rethink my evolutionary beliefs. Instead of jettisoning God, I instead needed to understand how a good and loving God – which is what Christians told me God is like – could allow so much pain, suffering, and death in this world. I did not need atheism, but rather theological understanding. I could see that molecules-to-man evolutionism was on shaky ground. I was questioning whether it was based on real science, or if it was just a bad philosophy. Once I began to question evolutionism, it did not take long before I found myself leaning toward the latter explanation.

Not everyone had that same experience, of course. Dan Brown, the author of the wildly popular book-turned-movie *The Da Vinci Code*, had a totally different experience from me. Brown revealed his unfortunate turn of events to Parade Magazine:

> I was raised Episcopalian, and I was very religious as a kid. Then, in eighth or ninth grade, I studied astronomy, cosmology, and the origins of the universe. I remember saying to a minister, "I don't get it. I read a book that said there was an explosion known as the Big Bang, but here it says God created heaven and

Earth and the animals in seven days. Which is right?" Unfortunately, the response I got was, "Nice boys don't ask that question." A light went off, and I said, "The Bible doesn't make sense. Science makes much more sense to me." And I just gravitated away from religion.[447]

Maybe Dan Brown was not open to exploring the 'big questions' of science and faith like I was. On the other hand, he might have been every bit as curious as I was, but maybe the right people never came into his life. Had knowledgeable and caring people come to his aid, there might never have been a book written by him – or maybe he would have written a Christian book instead. Only God knows how it could have worked out. But this example from Dan Brown's life demonstrates why apologetics is still needed in today's world.

Of course, Dan Brown's situation is nothing new. That is not surprising, right? How many times have you read that 'there is nothing new under the sun' (Ecclesiastes 1:9) in this book? The third century skeptic Porphyry, who was briefly mentioned in chapter two, experienced a similar situation to that of Dan Brown:

> Porphyry, who was known as the greatest enemy of Christianity, initially took a serious interest in Christianity in his youth and was intrigued by Christian leaders, like Origen of Alexandria (and later Caesarea). From what survives from his writings he knew his Bible better than many Christians today. It was only after a bad experience with Christians in Caesarea that Porphyry rejected a religion that produced badly behaving people. How different history could have been if only Christians in Caesarea had behaved differently![448]

This explains why Peter commands us to always talk to unbelievers in the spirit of 'gentleness and respect' (1 Peter 3:15). We should never do otherwise.

Despite the claim of some well-meaning Christians, apologetics is not dead yet. It can never be allowed to die, because we all have work to do: The 'apologetics mission field' is large indeed. Let us all get to work.

---

[447] James Kaplan, "Life After the Da Vinci Code." https://parade.com/106060/jameskaplan/13-dan-brown-life-after-da-vinci-code/

[448] Benno Zuiddam, "Battle for the Bible in the Early Church." https://creation.com/battle-for-bible-in-early-church

AFTERWORD

Reverend Lee Johnson
Retired Senior Pastor, Bethel Baptist Church in Galesburg, Illinois

If you are reading this 'Afterword,' having finished reading his book (as I assume you have), then you know that my friend Randy is a scholar. He is a thorough researcher and a man who honestly pursues the truth; those are two characteristics that any thinking person will appreciate. Those who read this book with an open mind will be forced to rethink their position on creation, whatever it is, because he has candidly covered all the bases. Interestingly enough, Randy and I approached the creation debate from the opposite poles of the spectrum, and then both of us, in turn, changed sides.

Randy became a Christian as an adult who just assumed to that point, like so many Americans, that Darwin got it right. He was taught in the public arena that the earth is four billion plus years old, and the universe is perhaps in the neighborhood of 13-15 billion years of age. He was taught that the biblical account of creation was ancient myth.

I, on the other hand, grew up reading and believing the marginal entry in my KJV Bible that 4004 BC was a date inspired by God. I assumed that Moses had specified exactly when God created the universe, and it was not to be questioned.

Randy has shared his story; my story went like this. If you believed the Bible was inspired by God then you had no option but to accept that Genesis 1:1-2:3 is to be taken as a literal, seven day, 24 hours per day series of creation events. If you thought otherwise, this meant that you did not accept that the Bible is infallible. The end result would be a mind opened to theological liberalism which would eventually not only cause you to deny God as the Creator, but it would lead you to reject every miracle in the Bible up to and including the physical resurrection of Jesus. It all made sense to me. Give up one inch and you will have lost the entire war.

My story also includes three years of a very biblical Bible college which brought me to love and respect the Bible as a book inspired by God. I took and passed every theology exam and honestly declared that I agreed with the position that the Bible is a book from God, without error in the original manuscripts, and the only rule for life and faith. I believed this.

After earning my college degree, I decided that God had called me to Christian ministry. I also knew that I needed a seminary level education so off I went, again to a quite conservative institution.

My theology on creation was not challenged in any of the seminary classes I took, but something happened in the final weeks before graduation that changed the course of my life. I took a class from a Chinese professor, a man whom I would describe as a language genius or perhaps even as an apostle, when I 'got it.' I have frequently said that in one instant of illumination my belief that the Bible was a book from God exploded. What I had mouthed for the great majority of my life about the Bible being a book from God changed from a theological proposition to be agreed with, and even to be argued over, to something that I 'knew to be true.' That the Bible is a book from God, the Creator of the heavens and the earth and the source of all things, I now saw to be an indisputable truth. How can I express that what I already believed I now came to BELIEVE! It changed my ministry, my preaching, my teaching, and certainly my life. But what did this have to do with Genesis and creation?

Over the next fifteen or twenty years I became increasingly uneasy with my position on the first chapter of Genesis as I read about new scientific discoveries. Then one day I was reading *The Confessions of Saint Augustine* when I ran across a statement on what a Christian should do if the Bible and science came into conflict.

I had no doubt that Augustine's answer would be to go with the Bible. You can perhaps appreciate the shock to my systematic theology when I read something to the effect that: "The Christian should see where his understanding of the Bible is wrong!" I was floored! How could a fourth to fifth century Christian genius theologian write those words? The end result was that it forced me to consider that perhaps I was reading Genesis 1:1-2:3 through the wrong set of lenses.

This is not the place to go into detail, so I will just say that I came to the conclusion that the problem lay in my view of the text. I decided that I was reading those verses as if they were a modern scientific treatise when in fact it was written as something very much different. It eventually led me to remaining a person who is unwaveringly convinced both that God is the Creator and that the Bible is a book inspired by God, but who also believes that the ages of both the universe and the earth are to be measured by billions rather than thousands of years. So, Randy and I disagree on that, yet he asked me to write this Afterword to his book. Why?

Two of the things I appreciate about Randy Hroziencik are his humility and his honesty when it comes to disputable matters like YEC vs. OEC. He understands the real issue: Is the material universe an accident, or was it designed and created by an omnipotent and omniscient God? Regarding this question, Randy and I have absolute agreement. On the secondary issue of YEC vs. OEC we do not agree, yet we have demonstrated that it is possible to disagree agreeably.

I would encourage you to learn two things from my friend Randy. First, the real battle is whether there is a Creator God or if everything is an incredible accident; that fight is non-negotiable. The second battle is how you will relate to someone who doesn't agree with your position on a disputable matter (Romans 14:1). If you know Randy, then you have learned from him that you will treat that person with charity, as Jesus taught.

So, I urge everyone to read the book and to consider Randy's arguments with care. In the end the reader will just have to come down on one side of the ledger or the other concerning the question of the age of the heavens and the earth. Wherever you land, both Randy and I will be on your side.

FOR FURTHER STUDY

The resources listed cover the major areas studied in this book. Many of the resources are from a recent creationist perspective, although I have included material from old earth creationists as well. I have not included material about atheism, pantheism, and deism since I addressed these worldviews in my previous book, *Worldviews in Collision: The Reasons for One Man's Journey from Skepticism to Christ*. The resources that I have listed are those I am familiar with. There are many other good resources as well, that a search on the internet will turn up. The resources listed here should be more than adequate to get one started on further study, however.

The reason I include resources from both young earth and old earth creationism is because a good theologian-apologist will never fear looking at both sides of the issue. Regardless of your persuasion, it will always be in your best interest to know what other Christians think about various issues. Doing so will make you a more well-rounded believer.

After much consideration, I decided to include only internet resources, to save the reader any additional cost. (Purchasing multiple books and DVD's can get expensive.) If you read not only this book, but also read through the internet papers listed here in addition to perusing the recommended websites, you will be busy for a long time! If you read through the bibliography you will find the names of books and videos that can be purchased for continued study as well. Additionally, I also recommended several books and DVD's throughout the text.

History of the Origins Debate

Jerry Bergman, "Darwinism and the Nazi Race Holocaust."
https://creation.com/darwinism-and-the-nazi-race-holocaust
Jerry Bergman, "Evolutionary Naturalism: An Ancient Idea."
https://creation.com/evolutionary-naturalism-an-ancient-idea
Danny Faulkner, "Heraclitus: Original Proponent of the Eternal Universe."
https://answersingenesis.org/astronomy/heraclitus-original-proponent-eternal-universe/
Michael Flannery, "What Did Darwin Know and When Did He Know It?"
https://reasons.org/explore/blogs/todays-new-reason-to-believe/read/tnrtb/2009/08/14/what-did-darwin-know-and-when-did-he-know-it
David Green, "The Long Story of Long Ages."
https://answersingenesis.org/genesis/the-long-story-of-long-ages/
Russell Grigg, "A Brief History of Design."
https://creation.com/a-brief-history-of-design
Russell Grigg, "Darwinism: It Was All in the Family."
https://creation.com/darwinism-it-was-all-in-the-family
Russell Grigg, "Darwin's Illegitimate Brainchild."

https://creation.com/charles-darwins-illegitimate-brainchild

Russell Grigg, "Ernst Haeckel: Evangelist for Evolution and Apostle of Deceit."
https://creation.com/ernst-haeckel-evangelist-for-evolution-and-apostle-of-deceit

Russell Grigg, "The Galileo Twist."
https://creation.com/the-galileo-twist

Russell Grigg, "Nietzsche: The Evolutionist Who was Anti-God and Anti-Darwin."
https://creation.com/nietzsche-anti-god-anti-darwin/

Paul James-Griffiths, "Evolution: An Ancient Pagan Idea."
https://creation.com/evolution-ancient-pagan-idea

G.J. Keane, "The Ideas of Teilhard de Chardin."
https://creation.com/teilhard-de-chardin

Louis Lavallee, "The Early Church Defended Creation Science."
https://icr.org/article/early-church-defended-creation-science

John Mackay & T. Parsons, "Pierre Simon Laplace: The Nebular Hypothesis."
https://answersingenesis.org/astronomy/cosmology/pierre-simon-laplace-the-nebular-hypothesis/

David Menton, "The Origin of Evolutionism: It Didn't Begin with Darwin."
https://answersingenesis.org/theory-of-evolution/origin-of-evolutionism-didnt-begin-with-darwin/

Henry Morris, "Evolution and the New Age."
https://icr.org/article/evolution-new-age/

Terry Mortenson, "Philosophical Naturalism and the Age of the Earth: Are They Related?"
https://answersingenesis.org/age-of-the-earth/are-philosophical-naturalism-and-age-of-the-earth-related/

Jonathan Sarfati, "The Biblical Roots of Modern Science."
https://creation.com/biblical-roots-of-modern-science

Andrew Sibley, "Adam as the Protoplast."
https://creation.com/adam-protoplast-or-archetype

Andrew Sibley, "*Bathybius haeckelii* and a 'Reign of Terror.'"
https://creation.com/bathybius-haeckelii

Andrew Sibley, "Deep Time in 18th-Century France – Part I: A Developing Belief."
https://creation.com/deep-time-in-18th-century-france-part-1

Andrew Sibley, "Deep Time in 18th-Century France – Part 2: Influence Upon Geology and Evolution in 18th and 19th Century Britain."
https://creation.com/deep-time-in-18th-century-france-part-2

Andrew Sibley, "Lessons from Augustine's *De Genesi ad Litteram-Libri Duodecim* (*The Literal Meaning of Genesis in Twelve Books*)."
https://creation.com/lessons-from-augustine

Andrew Sibley, "Orang-Outang or *Homo Sylvestris*: Ape-Men Before Darwin."
https://creation.com/apemen-before-darwin

Andrew Sibley, "Origen, Origins, and Allegory."
https://creation.com/origen-origins-and-allegory

Dominic Statham, "Christian Theology and the Rise of Newtonian Science – Imposed

Law and the Divine Will."
  https://creation.com/christianity-and-newtonian-science

Dominic Statham, "Darwin, Lyell and *Origin of Species*."
  https://creation.com/darwin-and-lyell

Dominic Statham, "The Truth About the Galileo Affair."
  https://creation.com/galileo-church

Lael Weinberger, "The Fall and the Inspiration for Science."
  https://creation.com/the-fall-inspiration-for-science

Lael Weinberger, "Reading the Bible and Understanding Nature."
  https://creation.com/review-harrison-bible-protestantism-natural-
    science

John Woodmorappe, "Hitler the Evolutionist; Hitler the Pantheist (Hitler the Atheist – Yes)."
  https://creation.com/review-hitlers-religion-weikart

Benno Zuiddam, "Augustine: Young Earth Creationist."
  https://creation.com/augustine-young-earth-creationist

Benno Zuiddam, "Battle for the Bible in the Early Church."
  https://creation.com/battle-for-bible-in-early-church

Benno Zuiddam, "Early Church Fathers on Creation, Death and Eschatology."
  https://creation.com/images/pdfs/tj/j28_1/j28_1_77-83.pdf

Benno Zuiddam, "Second Century Church Fathers: God Will Make Lions Vegetarian Again."
  https://creation.com/church-fathers-vegetarian-lions

Benno Zuiddam, "Was Evolution Invented by Greek Philosophers?"
  https://creation.com/
  https://www.creation.com/images/pdfs/tj/j32_1/j32_1_68-75.pdf

Refuting Evolutionism

Cosmological Evolutionism: Jason Lisle, "Does the Big Bang Fit with the Bible?"
  https://answersingenesis.org/big-bang/does-the-big-bang-fit-with-the-
    bible/

Geological Evolutionism: Steven Austin & Andrew Snelling, "Startling Evidence for Noah's Flood."
  https://creation.com/startling-evidence-for-noahs-flood

Biological Evolutionism: Monty White, "Hasn't Evolution Been Proven True?"
  https://answersingenesis.org/theory-of-evolution/evidence/hasnt-
    evolution-been-proven-true/

Ape-to-Man Evolutionism: Christopher Rupe & John Sanford, "Ape-Men or Just Ape and Man?"
  https://logosra.org/copy-of-spontaneous-life

Case for a Young Earth

Danny Faulkner, David Menton, Georgia Purdom, & Andrew Snelling, "Evidence for a Young Earth and Creation."

https://answersingenesis.org/creation-vs-evolution/evidence-for-young-earth-creation/

Terry Mortenson, "Philosophical Naturalism and the Age of the Earth: Are They Related?"
https://answersingenesis.org/age-of-the-earth/are-philosophical-naturalism-and-age-of-the-earth-related/

Case for an Old Earth

Jon Greene, "A Biblical Case for Old-Earth Creationism."
http://godandscience.org/youngearth/old_earth_creationism.html#when

Rodney Whitefield, "Genesis One and the Age of the Earth: What Does the Bible Say?"
https://creationingenesis.com/Genesis_One_and_the_Age_of_the_Earth.pdf

Suffering & Evil

Jonathan Sarfati, "Why Would a Loving God Allow Death & Suffering?"
https://creation.com/why-death-suffering

Rick Wade, "I Doubt the Existence of a Good God Who Allows a Baby to Suffer and Die."
https://probe.org/i-doubt-the-existence-of-a-good-god-who-allows-a-baby-to-suffer-and-die/

Eternal Fate of the Unevangelized

John Sanders, "Christian Approaches to the Salvation of Non-Christians."
http://drjohnsanders.com/christian-approaches-to-the-salvation-of-non-christians/

John Sanders, "Hell Yes! Hell No! Evangelical Debates on Eternal Punishment."
http://drjohnsanders.com/hell-yes-hell-no-evangelical-debates-on-eternal-punishment/

Heaven and the Afterlife

Dale Mason, "Let's Talk about Heaven."
https://answersingenesis.org/bible/lets-talk-about-heaven/

J. Warner Wallace, "Are There Any Good Reasons to Believe in Heaven (Even Without the Evidence from Scripture)?"
https://coldcasechristianity.com/writings/are-there-any-good-reasons-to-believe-in-heaven-even-without-the-evidence-from-scripture/

Case for Christ

William Lane Craig, "The Evidence for Jesus."
https://reasonablefaith.org/writings/popular-writings/jesus-of-nazareth/the-evidence-for-jesus/

Patrick Zukeran, "The Apologetics of Jesus: A Defense of His Deity."
https://probe.org/the-apologetics-of-jesus/

Recent Creationism Websites

Answers in Genesis (answersingenesis.org)
Biblical Science Institute (biblicalscienceinstitute.com)
Creation Astronomy (creationastronomy.com)
Creation Ministries International (creation.com)
Creation Today (creationtoday.org)
Institute for Creation Research (icr.org)
Logos Research Associates (logosra.org)

Old Earth Creationism Websites

God & Science (godandscience.org)
Old Earth Ministries (oldearth.org)
Reasons to Believe (reasons.org)

Theistic Evolutionism Websites

BioLogos (biologos.org)
Christianity & Evolution (theistic-evolution.com)
Talk Origins Archive (talkorigins.org)

General Apologetics Websites

Apologetics 315 (apologetics315.com)
Cold Case Christianity (coldcasechristianity.com)
Evidence for Christianity (evidenceforchristianity.org)
Gary Habermas (garyhabermas.com)
Probe Ministries (probe.org)
Reasonable Faith (reasonablefaith.org)

GLOSSARY OF TERMS

Although you may already be familiar with many of these terms, the following definitions may be helpful in this study on the origins debate. You do not want to miss a crucial point because of a misunderstanding in terminology. The terms related to evolutionism and creationism will be defined first, as they are the focus of this book. The other important terms will appear in alphabetical order after that.

Evolution

Evolution, in its most basic sense, means nothing more than change over time. In terms of origins, evolution consists of three aspects: Cosmological, geological, and biological. When most people think about evolution, they usually focus on biological change over time. However, it is important to remember that evolution in its 'big picture' sense encompasses three aspects: The universe in its entirety (cosmological evolution), our planet (geological evolution), and life here on the earth (biological evolution).

Biological Microevolution

Biological microevolution refers to small-scale change over time. (Micro means 'small.') Microevolution is change within a created 'kind' or, in biological terms, change within a species or even at the family level of classification. Microevolution is both scientific and scriptural. No one, evolutionist or creationist, should deny that small-scale changes occur within each distinct type (biblical kind) of plant or animal.

Biological Macroevolution

Biological macroevolution refers to large-scale change over time. (Macro means 'large.') Macroevolution is molecules-to-man progression over billions of years. The evolutionist claims that given enough microevolutionary change over time, an entirely new plant or animal will arise. However, macroevolution is unscriptural as plants and animals bring forth only according to their 'kind' (Genesis 1:11-12, 21, 24-25). It is also scientifically unproven and unsound, as there is no proof or even strong evidence that a single cell developed from non-living material, then diversified into every living thing from that cell. (In fact, the idea of the spontaneous generation of life from non-living matter has been thoroughly refuted throughout the past four centuries, beginning with Francesco Redi in the seventeenth century.) The origins debate is focused exclusively on macroevolutionary claims. Microevolution is not an area of debate concerning origins.

Evolutionism

Evolutionism – that is, evolution with the 'ism' ending – is the philosophical belief that the universe, our world, and life unfolded through natural processes over eons of time. There are three main versions of evolutionism. Atheistic evolutionism is change over time with no God or gods in the picture. Theistic evolutionism is change over time with an intelligence – usually the God of the Bible, but not always – guiding the process. Pantheistic evolutionism is the 'spiritual-but-impersonal' version of origins which maintains that everything emanates from the creative god-force behind the universe. All three versions of evolutionism are discussed in this book. Atheists, pantheists, deists, and theistic evolutionists adhere to the principles of evolutionism in some way, shape, or form. (Pantheism and deism are defined later in this section.)

At this point, let us review the terms related to evolution. Evolution simply refers to change over time. That could mean cosmological change over time, geological change over time, or biological change over time. Things do physically change over time, of course, so evolution is not an inherently evil term by any stretch of the imagination. Biologically, plants and animals – as well as people – can experience small-scale changes (microevolutionary change). A pair of dogs from the ancient past, each possessing a plethora of genetic material, can give rise to a wide variety of dogs in the future. (Although these future dogs will have lost that maximum genetic potential over time, lacking the ability to breed back to the original variety of dog with its rich genetic code.) Microevolution is scientific – biologists often refer to it as 'speciation' – and it is also scriptural as plants and animals bring forth according to their kinds (Genesis 1:11-12, 21, 24-25). So far, no controversy.

However, many scientists insist that, given enough microevolutionary changes over time, new plants and animals can arise. Evolutionists insist that not only did life originate from non-living material, but the first cell in turn gave rise to every type of plant or animal that has ever existed. This is the idea, or philosophy, known as macroevolution. There is no way that life could originate from non-life – this idea flies in the face of both Scripture and the Law of Biogenesis, which is a proven fact of science – and there is no clear-cut, undisputed series of fossils that link all the various plants and animals in the earth's history back to this original cell. Not even close. Microevolutionary change has limits. Small-scale changes over time, even if that could be over millions or billions of years, do not automatically lead to new life forms. Natural selection only serves to keep a species strong, not to change it into something completely different. In biology, stasis (biological consistency) seems to be the norm.

Evolutionism, with the 'ism' ending, denotes the philosophy related to evolution. Beginning with the so-called 'Big Bang,' the universe supposedly exploded into existence, from nothingness. Although a logical impossibility, this idea does not stop secular-minded scientists from accepting evolutionism. (We must ask ourselves the question, "Why is there something rather than nothing?" Without a Creator, there should be nothing – as there was prior to this supposed explosion.) Then, over billions of years the universe is believed to have formed into its present condition. Along the way, our solar system and planet formed through natural processes, and then life miraculously formed from non-life here on the earth. (How anyone ever gets past that obstacle is still a mystery to me.) Finally, through biological evolution here we are! This is nothing more than a humanistic philosophy dressed in scientific jargon.

Creationism

Creationism is the philosophical belief that God – usually, but not always, the God of the Bible – created the universe and everything in it. As with evolutionists, creationists come in various forms. Evolutionary creationists, who are also known as theistic evolutionists, believe that God created everything through continually guided evolutionary processes. In the context of this book, theistic evolutionists will be considered as evolutionists rather than creationists, although many of them would argue otherwise. (That is not a slam against theistic evolutionists personally. Rather, it is simply that I consider theistic evolution to be evolutionistic in nature, despite their oftentimes strong Christian convictions.) Old earth creationists believe that God created everything apart from evolutionary processes, but over eons of time. Young earth creationists (historic special creationists or recent creationists) also maintain that God created everything apart from evolutionary processes, but over the course of only six 24-hour days, and only thousands – not millions or billions – of years ago.

Agnosticism

Agnosticism comes from two Greek words, a, which means 'no' or 'without,' and gnosis, meaning 'knowledge.' Therefore, agnosticism describes the position in which its adherents possess 'no knowledge,' in this case concerning the existence of God. The agnostic is simply unsure regarding God's existence. Agnosticism comes in two varieties. Soft agnosticism is the position that people do not know if God exists, whereas hard agnosticism is the position that people cannot know if God exists. The soft agnostic believes that people currently do not know if God exists, but perhaps someday in the future there will be irrefutable evidence either for or against the existence of God. The hard agnostic believes that people will never have irrefutable evidence one way or the other. In their minds, God is unknowable, both now and in the future.

Apologetics

Apologetics is derived from the Greek word apologia, which means 'defense,' or more literally 'a speech for the defense.' Not surprisingly, apologetics is often associated with a legal-type defense of the faith. Apologia is found seven times in the New Testament (Acts 22:1; 25:16; 1 Corinthians 9:3; Philippians 1:7, 16; 2 Timothy 4:16; 1 Peter 3:15). When considering both the noun form apologia and the verb form apologeomai, the term appears seventeen times in the New

Testament. 1 Peter 3:15 is the verse which is generally held to be the 'battle cry' of Christian apologetics: "But in your hearts revere Christ as Lord. Always be prepared to give an answer to everyone who asks you to give the reason for the hope that you have. But do this with gentleness and respect."[449] Apologetics argues for various truth claims, such as the claim that God created the universe. It may also be used to demonstrate Christianity's power of interpretation regarding a variety of subjects, such as making a case for just war. Finally, for many people apologetics is all about refuting intellectual or emotional attacks against the Christian faith. An example of this would be establishing the case for the resurrection of Christ, as skeptics of Christianity are quick to deny this central tenet of the faith.

Atheism

Atheism comes from two Greek words, a, which means 'no' or 'without,' and theos, meaning 'God.' Therefore, atheism describes the philosophical position in which its adherents are 'without God.' The atheist is a person who does not believe in God, whether the God of the Bible or any other definition of a so-called 'Higher Power.' Likewise, the atheist denies the reality of anything supernatural, be it Heaven, Hell, angels, demons, or the existence of the immortal soul. In the atheistic worldview, only nature, or matter, exists. Therefore, atheism may also be referred to as either 'naturalism' or 'materialism.'

The Church

The Church, with a capital c, refers to the body of Christian believers worldwide. Regardless of the various branch or denomination a person may be affiliated with, all those who "declare with your mouth, 'Jesus is Lord,' and believe in your heart that God raised him from the dead"[450] are saved by Christ and together make up the Church universal. On the other hand, an individual congregation or local body of believers may be referred to as 'a church,' with a small-case c.

Creation *ex nihilo*

Creation *ex nihilo* is Latin for creation 'out of nothing.' Creation *ex nihilo* is summarized in the very first verse of Scripture: "In the beginning, God created the heavens and the earth."[451] The universe has not always existed, but rather God brought time, space, and matter-energy into existence at a finite point in the distant past.

---

[449] 1 Peter 3:15, NIV.
[450] Romans 10:9, NIV.
[451] Genesis 1:1, NIV.

Deism

Deism is a natural religion which recognizes God as Creator, but not as being intimately involved in the affairs of human beings. The god of deism – who is sometimes referred to as the 'First Cause,' the 'Grand Architect of the Universe,' or simply as 'Nature's God' – does not reveal himself/herself/itself through the written word. In this way, deism differentiates itself from the revealed religions of Judaism, Christianity, and Islam. Deists may be described as natural theologians who stress human reasoning, while at the same time rejecting special revelation of any kind, be it the Bible, the Qur'an, the Book of Mormon, and so forth. As would be expected, deists reject the concept of the Trinity, since this is a distinctly Christian concept, and instead view God as existing in the form of one person. Examples of deists throughout history include Voltaire, who is often mistakenly categorized as an atheist, and Thomas Paine. Although perhaps not 'hardcore' deists in the same vein as Voltaire and Paine, both Thomas Jefferson and Benjamin Franklin were strongly influenced by deism. The late Antony Flew, arguably the foremost atheist in the middle of the twentieth century, converted to deism in his later years.

The Early Church Fathers

The Early Church Fathers were those Christian leaders, theologians, and apologists of the earliest centuries of the Church who were wholly concerned with upholding the teachings of Christ and the biblical writers. The Apostolic Fathers lived in the first and second centuries, and they personally knew one or more of the Twelve Apostles. The Apostolic Fathers would include men such as Clement of Rome, Ignatius of Antioch, and Polycarp of Smyrna. After the Apostolic Fathers came many great leaders such as Justin Martyr, Irenaeus, Tertullian, Clement of Alexandria, Origen, Augustine, and too many other names to mention. Collectively, the Apostolic Fathers and those great leaders which came shortly after them are referred to as the 'Early Church Fathers.'

General Revelation

General revelation is God's self-disclosure of himself through the created order and the inherent moral consciousness in all people. Scriptural verses which describe God's general revelation through nature include Psalm 19:1, Acts 14:16-17, and Romans 1:20, while Romans 2:14-15 describes the innate moral consciousness of all human beings. General revelation, unlike special revelation (Scripture), has been given to all people throughout all time. Therefore, all people who have ever lived possessed the ability to comprehend God's existence.

Intelligent Design

Intelligent design is the belief that nature exhibits patterns that are best explained as the products of intelligence rather than random chance. Intelligent design may be summarized in the popular phrase, "An intelligent design demands an intelligent Designer." Proponents of the Intelligent Design Movement (ID) stop short of declaring that the 'intelligent Designer' is none other than the God of the Bible. Instead, they are content to simply point out that Darwinism is insufficient to explain the origin and diversity of life in the world today, as well as in the fossil record. ID utilizes a variety of academic disciplines to make a comprehensive case for a Designer – from astrophysics to microbiology, and everything in between. No stone is left unturned when pointing out the erroneous assumptions that are being made in the naturalistic view of origins. One of the amazing things about ID is that it involves many scientists and scholars from a variety of fields, and from a variety of religious backgrounds – or from no religious background at all. Since so many of the scientists and scholars in ID are from diverse backgrounds and are united simply in the belief in an intelligent Designer of some sort, many skeptics in recent times have been persuaded to give design theory a second look.

Monism

Monism is the philosophical idea that everything in the universe is of one substance, and everything is connected to everything else.

Paganism

Paganism in the Greco-Roman world referred to all religions other than the Jews and Christians. "The term pagan comes from the Latin pagus, meaning 'the countryside,' and pagani, meaning 'rural people.' The term seems to have been applied to followers of the old [polytheistic] religions when Christianity became a dominant religious force within the Empire. Since Christianity was primarily an urban religion, followers of the older religions mostly lived in the countryside, and since they were pagani (rural people), their religions were described as 'pagan.'"[452] Today, this may be used as a pejorative term by some, but in the context of this book paganism refers to non-biblical, polytheistic religions and philosophies which are in opposition to biblical truth. Some religious groups today refer to themselves as 'neo-pagan.'

Pantheism

Pantheism comes from two Greek words, pan, meaning 'all,' and theos, meaning 'God.' Therefore, the pantheist is one who believes that 'all is God.' In short, pantheism teaches that everything is God, and everything is of one substance (monism). Pantheism is often referred to as 'pantheistic monism' or 'monistic pantheism,' as the two concepts are often inseparably joined. Pantheists make the mistake of confusing the creation with the Creator, a situation which is addressed by Paul in Romans 1:25. Many ancient Greek

---

[452] Glenn Sunshine, *Why You Think the Way You Do* (Grand Rapids, MI: Zondervan, 2009), 20-21.

philosophers were pantheists. Eastern religions in general also adhere to pantheism. This includes Hinduism, some forms of Buddhism, and Taoism. New Age spirituality, which heavily incorporates Eastern religious concepts, is based in large part upon pantheism. Even some self-proclaimed Christian groups, such as Unity School of Christianity and Christian Science, incorporate the pantheistic worldview into their doctrines.

Polytheism

Polytheism comes from two Greek words, poly, meaning 'many,' and theos, meaning 'God.' Therefore, the polytheist is one who believes in many gods. Polytheism stretches back to the earliest civilizations: Sumer, Egypt, Babylon, and through to the Greco-Roman world and beyond. Today, polytheism remains very much alive-and-well among religious groups such as the Mormons, Hindu's, and neo-pagans of all stripes. Generally, polytheistic cultures tended to worship the various gods out of fear, hoping to appease the gods in the hope that they would, at best, bring good things and, at worst, simply leave them alone. The Israelites were surrounded by polytheistic peoples throughout their ancient history: When Moses wrote down the Genesis creation account, he almost certainly had that thought in mind.

Scripture

Scripture refers to the Bible. Although many religions have their own version of the scriptures, only the Bible is God's true revelation to mankind for the Christian believer. Scripture includes the Old and New Testaments, from Genesis to Revelation.

Skepticism

Skepticism comes from the Latin word scepticus, which means 'inquiring,' 'reflective,' or 'doubtful.' Scepticus in turn is derived from the Greek word scepsis, meaning the same thing but also with an emphasis upon 'hesitation.' The religious skeptic is one who is inquisitive concerning matters of faith, yet at the same time is doubtful that humanity can answer the 'big questions of life.'

Special Revelation

Special revelation is God's self-disclosure of himself in the written word. The written word is, of course, the Bible. The beauty of Scripture is that we get a much more detailed look at both God's nature and his plan for humanity than we can receive from general revelation alone. With general revelation we can clearly see that God exists, and that he is awesome in his power to create, yet we can plainly see that things are not as they should be. (The world is 'messed up,' for lack of a better term.) We clearly live in a created world, but not everything is wonderful and good. Something is wrong, and we are all too aware of it. Scripture tells us why that is: This world was created by God, but it is now in a fallen state resulting from man's rebellion. Scripture gives us much more information about God and all things spiritual than general revelation alone could ever provide.

Theism

Theism, at least as defined in about every dictionary known to mankind, is the belief in either God or the gods. However, a belief in the gods is polytheism, which is markedly different from just 'theism.' Also, theism is different from pantheism in that the Creator-God of theism possesses not only the attributes of personality (not just a force), but also has a personal relationship with his creatures – especially mankind. A synonymous term for theism is 'monotheism,' which further serves to distinguish theism from both polytheism and pantheism. Examples of theists would be Jews, Christians, and Muslims.

Worldview

Worldview may be defined in several ways, but generally worldview is the lens through which we see all of reality. Worldview is how we make sense of the world in which we live. The atheistic worldview sees the world one way, the Christian worldview sees it another way, the pantheistic worldview differently still, and so on. Worldview tackles the big questions of life, such as: Does God exist, and if so, what is God like? Where is the basis of morality and values to be found? Who are we as human beings? What happens after we die? Although people are sometimes inconsistent in their worldview, in general most rational people attempt to be at least somewhat logically consistent in answering the big questions of life.

BIBLIOGRAPHY

Arnobius. *Against the Heathen, Book 1.*
https://newadvent.org/fathers/06311.htm

Arnobius. *Against the Heathen, Book 2.*
https://newadvent.org/fathers/06312.htm

Augustine. *The Confessions, Book 1.*
https://newadvent.org/fathers/110101.htm

Augustine. *The Literal Meaning of Genesis, Book 1.*
http://inters.org/augustine-interpretating-sacred-scripture

Ayala, Francisco. "Nothing in Biology Makes Sense Except in the Light of Evolution: Theodosius Dobzhansky, 1900-1975." Journal of Heredity (Volume 68, Number 3, 1977).

Bates, Gary. "Designed by Aliens? Discoverers of DNA's Structure Attack Christianity."
https://creation.com/designed-by-aliens-crick-watson-atheism-panspermia

Beebe, James. "The Kalam Cosmological Argument for the Existence of God."
http://apollos.squarespace.com/cosmological-argument/

Bell, Philip. *Evolution and the Christian Faith.* Leominster, England: Day One Publications, 2018.

Bergman, Jerry. "Evolutionary Naturalism: An Ancient Idea."
https://creation.com/evolutionary-naturalism-an-ancient-idea

Boyd, Greg. "The Case for Annihilationism."
https://reknew.org/2008/01/the-case-for-annihilationism/

Bray, Gerald. "Augustine's Key."
https://christianitytoday.com/history/issues/issue-80/augustines-key.html

Carter, Robert & Chris Hardy. "The Biblical Minimum and Maximum Age of the Earth."
https://creation.com/biblical-age-of-the-earth

Chapman, Allan. Slaying the Dragons: Destroying Myths in the History of Science and Faith. Oxford, England: Lion Books, 2013.

Cooper, Bill. *After the Flood.* West Sussex, England: New Wine Press, 1995.

Copan, Paul. *That's Just Your Interpretation.* Grand Rapids, MI: Baker Books, 2001.

Crowe, Donald. *Creation Without Compromise.* Brisbane, Australia: Creation Ministries International, 2009.

De Chardin, Teilhard. *The Phenomenon of Man.* London, England: William Collins Sons & Co. Ltd., 1980.

Eddy, John. Geotimes Magazine (September 1978). Report on Symposium at Louisiana State University.

Faulkner, Danny. "Heraclitus: Original Proponent of the Eternal Universe."
https://answersingenesis.org/astronomy/heraclitus-original-proponent-eternal-universe/

Faulkner, Danny & Bodie Hodge. "What About Distant Starlight Models?"
https://answersingenesis.org/astronomy/starlight/what-about-distant-starlight-models/

Flannery, Michael. "What Did Darwin Know and When Did He Know It?" https://reasons.org/explore/blogs/todays-new-reason-to-believe/read/tnrtb/2009/08/14/what-did-darwin-know-and-when-did-he-know-it

Geisler, Norman. *If God, Why Evil?* Bloomington, MN: Bethany House Publishers, 2011.

Geisler, Norman & Frank Turek. *I Don't Have Enough Faith to Be an Atheist*. Wheaton, IL: Crossway Books, 2004.

Gibbs, W. Wayt. "Profile: George F.R. Ellis: Thinking Globally, Acting Universally." Scientific American, Volume 273: Number 4 (October 1995).

Gould, Stephen. "The Late Birth of a Flat Earth" in *Dinosaur in a Haystack: Reflections in Natural History*. New York, NY: Three Rivers Press, 1997.

Green, Arthur Francis. "When Fables Fall." Unmasking Fables, Promoting Truth DVD (Creation Ministries International). https://usstore.creation.com/product/401-unmasking-fables-promoting-truth-dvd

Green, David. "The Long Story of Long Ages." https://answersingenesis.org/genesis/the-long-story-of-long-ages/

Greene, Jon. "A Biblical Case for Old-Earth Creationism." http://godandscience.org/youngearth/old_earth_creationism.html#when

Grigg, Russell. "A Brief History of Design." https://creation.com/a-brief-history-of-design

Grigg, Russell. "Darwinism: It Was All in the Family." https://creation.com/darwinism-it-was-all-in-the-family

Grigg, Russell. "Darwin's Illegitimate Brainchild." https://creation.com/charles-darwins-illegitimate-brainchild

Grigg, Russell. "Ernst Haeckel: Evangelist for Evolution and Apostle of Deceit." https://creation.com/ernst-haeckel-evangelist-for-evolution-and-apostle-of-deceit

Grigg, Russell. "The Galileo Twist." https://creation.com/the-galileo-twist

Grigg, Russell. "Nietzsche: The Evolutionist Who was Anti-God and Anti-Darwin." https://creation.com/nietzsche-anti-god-anti-darwin/

Guillen, Michael. *Amazing Truths: How Science and the Bible Agree*. Grand Rapids, MI: Zondervan, 2015.

Ham, Ken. "Evangelism for the New Millenium." https://answersingenesis.org/gospel/evangelism/evangelism-for-the-new-millennium/

Herbert, David. *The Faces of Origins*. London, Ontario: D & I Herbert Publishing, 2004.

Hoffman, Nathan. *Were the Pyramids Built Before the Flood?* https://www.youtube.com/watch?v=VI1yRTC6kGE&list=PL6P6ysO2XSKIcu--7aV8fB_78JrTKL_aI&index=6

Hroziencik, Randall. Worldviews in Collision: The Reasons for One Man's Journey from Skepticism to Christ. Greenwood, WI: Paley, Whately, and Greenleaf Press, 2018.

Huffling, Brian. "Apologetic Methods and a Case for Classical Apologetics." https://ses.edu/apologetic-methods-and-a-case-for-classical-apologetics/

James-Griffiths, Paul. "Evolution: An Ancient Pagan Idea."
> https://creation.com/evolution-ancient-pagan-idea

James-Griffiths, Paul. "Exposing the Roots of Evolution." Unmasking Fables, Promoting Truth DVD (Creation Ministries International).
> https://usstore.creation.com/product/401-unmasking-fables-promoting-truth-dvd

Jeffrey, Grant. *Journey into Eternity: Search for Immortality*. Toronto, Ontario: Frontier Research Publications, Inc., 2000.

Justin Martyr. *Second Apology*.
> https://newadvent.org/fathers/0127.htm

Kaplan, James. "Life After the Da Vinci Code."
> https://parade.com/106060/jameskaplan/13-dan-brown-life-after-da-vinci-code/

Keane, G.J. "The Ideas of Teilhard de Chardin."
> https://creation.com/teilhard-de-chardin

Lavallee, Louis. "The Early Church Defended Creation Science."
> https://icr.org/article/early-church-defended-creation-science

Lewis, C.S. *Mere Christianity*. San Francisco, CA: HarperSanFrancisco, 2001. 1952 Original Publication.

Mackay, John & T. Parsons. "Pierre Simon Laplace: The Nebular Hypothesis."
> https://answersingenesis.org/astronomy/cosmology/pierre-simon-laplace-the-nebular-hypothesis/

McDermott, Gerald. *God's Rivals*. Downers Grove, IL: InterVarsity Press, 2007.

McDowell, Sean & Jonathan Morrow. *Is God Just a Human Invention?* Grand Rapids, MI: Kregel Publications, 2010.

Meister, Chad. *Building Belief*. Grand Rapids, MI: Baker Books, 2006.

Menton, David. "The Origin of Evolutionism: It Didn't Begin with Darwin."
> https://answersingenesis.org/theory-of-evolution/origin-of-evolutionism-didnt-begin-with-darwin/

Menton, David. "The Origin of Life."
> https://answersingenesis.org/origin-of-life/origin-of-life-chance-events/

Morris, Henry. "Evolution and the New Age."
> https://icr.org/article/evolution-new-age/

Morris, Henry. *The Long War Against God*. Green Forest, AR: Master Books, 2000.

Morris, Henry. "Pantheistic Evolution."
> https://icr.org/article/pantheistic-evolution/

Mortenson, Terry. "Understanding Genesis 1 Hebrew: Create (*bara*) & Make (*asah*)."
> https://answersingenesis.org/genesis/did-god-create-bara-or-make-asah-in-genesis-1/

Muncaster, Ralph. *Dismantling Evolution: Building the Case for Intelligent Design*. Eugene, OR: Harvest House Publishers, 2003.

Oakes, John. *Reasons for Belief: A Handbook of Christian Evidence*. Spring, TX: Illumination Publishers International, 2005.

Osborn, Henry Fairfield. *From the Greeks to Darwin*. New York, NY: Charles Scribner's Sons, 1929.

Ovid. *Metamorphoses, Book 1*.
> https://ovid.lib.virginia.edu/trans/Metamorph.htm#488381088

Pascal, Blaise. *Pensees*. New York, NY; Penguin Books, 1966.

Rhodes, Ron. Answering the Objections of Atheists, Agnostics, & Skeptics. Eugene, OR: Harvest House Publishers, 2006.

Riddle, Mike. "Does Radiometric Dating Prove the Earth Is Old?" https://answersingenesis.org/geology/radiometric-dating/does-radiometric-dating-prove-the-earth-is-old/

Riddlebarger, Kim. "For the Sake of the Gospel: Paul's Apologetic Speeches." Modern Reformation. https://modernreformation.org/default.php?page=articledisplay&var1=ArtRead&var2=445&var3=main

Ross, Hugh. *The Creator and the Cosmos*. Colorado Springs, CO: NavPress Publishing Group, 1993.

Ross, Hugh. *The Fingerprint of God*. New Kensington, PA: Whitaker House, 1989.

Sanders, John. "Those Who Have Never Heard: A Survey of the Major Positions." https://rsc.byu.edu/salvation-christ-comparative-christian-views/those-who-have-never-heard-survey-major-positions

Sarfati, Jonathan. "The Biblical Roots of Modern Science." https://creation.com/biblical-roots-of-modern-science

Sarfati, Jonathan. "DNA: Marvelous Messages or Mostly Mess?" https://creation.com/dna-marvelous-messages-or-mostly-mess

Sarfati, Jonathan. *Refuting Compromise*. Green Forest, AR: Master Books, 2004.

Sarfati, Jonathan & Carl Wieland. "Culture Wars: Bacon vs Ham (Part 1)." https://creation.com/part-1-culture-wars-bacon-vs-ham

Schroeder, Gerald. "The Age of the Universe." http://geraldschroeder.com/AgeUniverse.aspx

Sibley, Andrew. *Cracking the Darwin Code*. Colyton, Devon: Fastnet Publications, 2013.

Sibley, Andrew. "Deep Time in 18th-Century France – Part I: A Developing Belief." https://creation.com/deep-time-in-18th-century-france-part-1

Sibley, Andrew. "Orang-Outang or *Homo Sylvestris*: Ape-Men Before Darwin." https://creation.com/apemen-before-darwin

Snelling, Andrew. "Radiometric Dating: Problems with the Assumptions." https://answersingenesis.org/geology/radiometric-dating/radiometric-dating-problems-with-the-assumptions/

Snoke, David. *A Biblical Case for an Old Earth*. Grand Rapids, MI: Baker Books, 2006.

Sproul, R.C. *The Consequences of Ideas*. Wheaton, Illinois: Crossway Books, 2000.

Sproul, R.C. *Reason to Believe*. Grand Rapids, MI: Zondervan, 1978.

Statham, Dominic. "The Truth About the Galileo Affair." https://creation.com/galileo-church

Stoner, Don. "The Historical Context for the Book of Genesis," Revision 2011-06-06, Part 3: Identifying Noah and the Great Flood. http://dstoner.net/Genesis_Context/Context.html#part3

Story, Dan. "How Does Secular Science Attempt to Disprove God's Existence? (Part Four)." http://danstory.net/blog/

Strobel, Lee. *The Case for Faith*. Grand Rapids, MI: Zondervan, 2000.

Sunshine, Glenn. *Why You Think the Way You Do*. Grand Rapids, MI: Zondervan, 2009.

Theophilus. *To Autolycus, Book 2*.
https://newadvent.org/fathers/02042.htm
Theophilus. *To Autolycus, Book 3*.
https://newadvent.org/fathers/02043.htm
Thomsen, Dietrick. "A Knowing Universe Seeking to be Known." Science News (Volume 123, February 19, 1983).
Turpin, Simon. "How Do Some Among You Say There Is No Adam?"
https://answersingenesis.org/adam-and-eve/how-do-some-among-you-say-there-no-adam/
Upchurch, John. "The Danger of BioLogos."
https://answersingenesis.org/theistic-evolution/the-danger-of-biologos/
Vince, Ron. "At the Areopagus (Acts 17:22-31): Pauline Apologetics and Lucan Rhetoric."
https://mcmaster.ca/mjtm/4-5.htm
Wiens, Roger. "Radiometric Dating: A Christian Perspective."
https://d4bge0zxg5qba.cloudfront.net/files/articles/non-staff-papers/roger_wiens_radiometric_dating.pdf/
Woodmorappe, John. "Hitler the Evolutionist; Hitler the Pantheist (Hitler the Atheist – Yes)."
https://creation.com/review-hitlers-religion-weikart
Zuiddam, Benno. "Augustine: Young Earth Creationist."
https://creation.com/augustine-young-earth-creationist
Zuiddam, Benno. "Battle for the Bible in the Early Church."
https://creation.com/battle-for-bible-in-early-church
Zuiddam, Benno. "Was Evolution Invented by Greek Philosophers?"
https://creation.com/
https://www.creation.com/images/pdfs/tj/j32_1/j32_1_68-75.pdf

ABOUT THE AUTHOR

Randall Hroziencik, a radiation therapist by profession, has a passion for teaching creationism, theology, and biblical studies. He has a Bachelor of Science in Health Arts from the University of Saint Francis (Joliet, Illinois) and a Master of Arts-Doctor of Philosophy in Theology from Trinity College of the Bible & Trinity Theological Seminary (Newburgh, Indiana). He was ordained through his home church, Bethel Baptist Church in Galesburg, Illinois and is the first-ever graduate of the Apologetics Research Society's Certificate in Christian Apologetics. He is also the author of *Worldviews in Collision: The Reasons for One Man's Journey from Skepticism to Christ*. He especially enjoys the history of the origins debate and how the various world religions and philosophies compare to the teachings of the Bible.

When not working or writing, he can be found bicycling, exercising, motorcycling, and reading. His greatest joy is his family. He and his wife, Debbie, are blessed to have both children – and their four granddaughters – living within just miles of them.

You may contact him at randyhroziencik@comcast.net